EDUCATION IN A COMPETITIVE AND GLOBALIZING WORLD

STRATEGIES FOR EFFECTIVE SCIENCE, TECHNOLOGY, ENGINEERING AND MATH (STEM) EDUCATION

Education in a Competitive and Globalizing World

Additional books in this series can be found on Nova's website under the Series tab.

Additional E-books in this series can be found on Nova's website under the E-book tab.

EDUCATION IN A COMPETITIVE AND GLOBALIZING WORLD

STRATEGIES FOR EFFECTIVE SCIENCE, TECHNOLOGY, ENGINEERING AND MATH (STEM) EDUCATION

VALERIAN I. BELL
AND
SHIOBAN E. GONZALEZ
EDITORS

Nova Science Publishers, Inc.
New York

For permission to use material from this book please contact us:
Telephone 631-231-7269; Fax 631-231-8175
Web Site: http://www.novapublishers.com

Additional color graphics may be available in the e-book version of this book.

LIBRARY OF CONGRESS CATALOGING-IN-PUBLICATION DATA

ISBN: 978-1-62081-523-6

Published by Nova Science Publishers, Inc. † New York

CONTENTS

PREFACE

Economic projections point to a need for approximately 1 million more STEM professionals than the U.S. will produce at the current rate over the next decade if the country is to retain its historical preeminence in science and technology. To meet this goal, the United States will need to increase the number of students who receive undergraduate STEM degrees by about 34% annually over current rates. This book provides an overview for strategies to improve STEM education during the first two years of college, thereby being responsive to both the challenges and the opportunities that this crucial stage in the STEM education pathway presents.

In: Strategies for Effective Science, Technology, Engineering... ISBN: 978-1-62081-523-6
Editors: Valerian I. Bell and Shioban E. Gonzalez

Chapter 1

ENGAGE TO EXCEL: PRODUCING ONE MILLION ADDITIONAL COLLEGE GRADUATES WITH DEGREES IN SCIENCE, TECHNOLOGY, ENGINEERING, AND MATHEMATICS*

Executive Office of the President, President's Council of Advisors on Science and Technology

ABOUT THE PRESIDENT'S COUNCIL OF ADVISORS ON SCIENCE AND TECHNOLOGY

The President's Council of Advisors on Science and Technology (PCAST) is an advisory group of the nation's leading scientists and engineers, appointed by the President to augment the science and technology advice available to him from inside the White House and from cabinet departments and other Federal agencies. PCAST is consulted about and often makes policy recommendations concerning the full range of issues where understandings from the domains of science, technology, and innovation bear potentially on the policy choices before the President. For more information about PCAST, see www.whitehouse.gov/ostp/pcast.

THE PRESIDENT'S COUNCIL OF ADVISORS ON SCIENCE AND TECHNOLOGY

Co-Chairs
John P. Holdren
Assistant to the President for Science and Technology
Director, Office of Science and Technology Policy

* This is an edited, reformatted and augmented version of Executive Office of the President, President's Council of Advisors on Science and Technology, Publication, dated February 2 012.

Vice Chairs
William Press
Raymer Professor in Computer Science and Integrative Biology
University of Texas at Austin

Members
Rosina Bierbaum
Professor of Natural Resources and Environmental
Policy School of Natural Resources and
Environment and School of Public Health
University of Michigan

Christine Cassel President and CEO
American Board of Internal Medicine

Christopher Chyba
Professor, Astrophysical Sciences and International
Affairs Director, Program on Science and Global
Security Princeton University

S. James Gates, Jr.
John S. Toll Professor of Physics
Director, Center for String and Particle Theory
University of Maryland, College Park

Mark Gorenberg
Managing Director
Hummer Winblad Venture Partners

Eric Lander
President
Broad Institute of Harvard and MIT

Maxine Savitz
Vice President
National Academy of Engineering

Shirley Ann Jackson
President
Rensselaer Polytechnic Institute

Richard C. Levin
President
Yale University

Chad Mirkin
Rathmann Professor, Chemistry,
Materials Science and Engineering,
Chemical and Biological Engineering
and Medicine Director, International Institute
for Nanotechnology Northwestern University

Mario Molina
Professor, Chemistry and Biochemistry University of California,
San Diego
Professor, Center for Atmospheric Sciences at the Scripps
Institution of Oceanography
Director, Mario Molina Center for Energy and Environment, Mexico City

Ernest J. Moniz
Cecil and Ida Green Professor of Physics and Engineering Systems
Director, MIT's Energy Initiative
Massachusetts Institute of Technology

Craig Mundie
Chief Research and Strategy
Officer Microsoft Corporation

Ed Penhoet
Director, Alta Partners
Professor Emeritus, Biochemistry
and Public
Health
University of California, Berkeley

Barbara Schaal
Mary-Dell Chilton Distinguished Professor of
Biology
Washington University, St. Louis
Vice President, National Academy of Sciences

Eric Schmidt
Executive Chairman Google, Inc.

Daniel Schrag
Sturgis Hooper Professor of Geology Professor,
Environmental Science and Engineering
Director, Harvard University
Center for the Environment
Harvard University

David E. Shaw
Chief Scientist, D.E. Shaw Research
Senior Research Fellow, Center for
Computational Biology and Bioinformatics
Columbia University

Ahmed Zewail
Linus Pauling Professor of Chemistry
and Physics Director,
Physical Biology Center
California Institute of Technology

Staff

Deborah Stine
Executive Director

Danielle Evers
AAAS Science and Technology Policy Fellow

Amber Hartman Scholz
Assistant Executive Director

EXECUTIVE OFFICE OF THE PRESIDENT PRESIDENT'S COUNCIL OF ADVISORS ON SCIENCE AND TECHNOLOGY, WASHINGTON, D.C. 20502

President Barack Obama
The White House
Washington, D.C. 20502

Dear Mr. President,

We are pleased to present you with this report, *Engage to Excel: Producing One Million Additional College Graduates with Degrees in Science, Technology, Engineering, and Mathematics*, prepared for you by the President's Council of Advisors on Science and Technology (PCAST). This report provides a strategy for improving STEM education during the first two years of college that we believe is responsive to both the challenges and the opportunities that this crucial stage in the STEM education pathway presents.

In preparing this report, PCAST assembled a Working Group of experts in postsecondary STEM teaching, learning-science research, curriculum development, higher-education administration, faculty training, educational technology, and successful interaction between industry and higher education. The report was strengthened by input from additional experts in postsecondary STEM education, STEM practitioners, professional societies, private companies, educators, and Federal education officials.

PCAST found that economic forecasts point to a need for producing, over the next decade, approximately 1 million more college graduates in STEM fields than expected under current assumptions. Fewer than 40% of students who enter college intending to major in a STEM field complete a STEM degree. Merely increasing the retention of STEM majors from 40% to 50% would generate three-quarters of the targeted 1 million additional STEM degrees over the next decade.

PCAST identified five overarching recommendations that it believes can achieve this goal: (1) catalyze widespread adoption of empirically validated teaching practices; (2) advocate and provide support for replacing standard laboratory courses with discovery-based research courses; (3) launch a national experiment in postsecondary mathematics education to address the mathematics-preparation gap; (4) encourage partnerships among stakeholders to diversify pathways to STEM careers; and (5) create a Presidential Council on STEM Education with leadership from the academic and business communities to provide strategic leadership for transformative and sustainable change in STEM undergraduate education.

Implementing these recommendations will help you achieve one of the key STEM goals you stated in your address to the National Academy of Sciences in April 2009: "American students will move from the middle to the top of the pack in science and math over the next decade. For we know that the nation that out-educates us today—will out-compete us tomorrow." The members of PCAST are grateful for the opportunity to provide our input on an issue of such critical importance to the Nation's future.

Sincerely,
John P. Holdren

PCAST Co-Chair

Eric Lander
PCAST Co-Chair

Executive Report

Economic projections point to a need for approximately 1 million more STEM professionals than the U.S. will produce at the current rate over the next decade if the country is to retain its historical preeminence in science and technology.

To meet this goal, the United States will need to increase the number of students who receive undergraduate STEM degrees by about 34% annually over current rates.

Currently the United States graduates about 300,000 bachelor and associate degrees in STEM fields annually. Fewer than 40% of students who enter college intending to major in a STEM field complete a STEM degree. Increasing the retention of STEM majors from 40% to 50% would, alone, generate three-quarters of the targeted 1 million additional STEM degrees over the next decade.

Many of those who abandon STEM majors perform well in their introductory courses and would make valuable additions to the STEM workforce. Retaining more students in STEM majors is the lowest-cost, fastest policy option to providing the STEM professionals that the nation needs for economic and societal well-being, and will not require expanding the number or size of introductory courses, which are constrained by space and resources at many colleges and universities.

The reasons students give for abandoning STEM majors point to the retention strategies that are needed. For example, high-performing students frequently cite uninspiring introductory courses as a factor in their choice to switch majors.

And low-performing students with a high interest and aptitude in STEM careers often have difficulty with the math required in introductory STEM courses with little help provided by their universities. Moreover, many students, and particularly members of groups underrepresented in STEM fields, cite an unwelcoming atmosphere from faculty in STEM courses as a reason for their departure.

Better teaching methods are needed by university faculty to make courses more inspiring, provide more help to students facing mathematical challenges, and to create an atmosphere of a community of STEM learners. Traditional teaching methods have trained many STEM professionals, including most of the current STEM workforce. But a large and growing body of research indicates that STEM education can be substantially improved through a diversification of teaching methods.

These data show that evidence-based teaching methods are more effective in reaching all students—especially the "underrepresented majority"—the women and members of minority groups who now constitute approximately 70% of college students while being underrepresented among students who receive undergraduate STEM degrees (approximately 45%). This underrepresented majority is a large potential source of STEM professionals.

The Need for an Improved STEM Student Recruitment and Retention Strategy for the First Two Years of Postsecondary Education

The first two years of college are the most critical to the retention and recruitment of STEM majors. These two years are also a shared feature of all types of 2- and 4-year colleges and universities—community colleges, comprehensive universities, liberal arts colleges, research universities, and minority-serving institutions. In addition, STEM courses during the first two years of college have an enormous effect on the knowledge, skills, and attitudes of future K-12 teachers. For these reasons, this report focuses on actions that will influence the quality of STEM education in the first two years of college.

Based on extensive research about students'choices, learning processes, and preparation, three imperatives underpin this report:

- Improve the first two years of STEM education in college.
- Provide all students with the tools to excel.
- Diversify pathways to STEM degrees.

Our recommendations, described below, detail how to convert these imperatives into action. The title of this report, "Engage to Excel," applies to students, faculty, and leaders in academia, industry, and government. Students must be engaged to excel in STEM fields. To excel as teachers, faculty must engage in methods of teaching grounded in research about why students excel and persist in college. Moreover, success depends on the engagement by great leadership. Leaders, including the President of the United States; college, university and business leadership; and others, must encourage and support the creation of well-aligned incentives for transforming and sustaining STEM learning. They also must encourage and support the establishment of broad-based reliable metrics to measure outcomes in an ongoing cycle of improvement. Transforming STEM education in U.S. colleges and universities is a daunting challenge. The key barriers involve faculty awareness and performance, reward and incentive systems, and traditions in higher education. The recommendations in this report address the most significant barriers and use both tangible resources and persuasion to inspire and catalyze change. Attacking the issue from numerous angles and with various tools is aimed at reaching a point at which the movement will take on a momentum of its own and produce sweeping change that is sustainable without further Federal intervention.

Recommendations

The President's Council of Advisors on Science and Technology (PCAST) proposes five overarching recommendations to transform undergraduate STEM education during the transition from high school to college and during the first two years of undergraduate STEM education:

1) Catalyze widespread adoption of empirically validated teaching practices.
2) Advocate and provide support for replacing standard laboratory courses with discovery-based research courses.
3) Launch a national experiment in postsecondary mathematics education to address the math preparation gap.
4) Encourage partnerships among stakeholders to diversify pathways to STEM careers.
5) Create a Presidential Council on STEM Education with leadership from the academic and business communities to provide strategic leadership for transformative and sustainable change in STEM undergraduate education.

Each of these recommendations will be explained in more detail below.

Recommendation 1

Catalyze Widespread Adoption of Empirically Validated Teaching Practices

Learning theory, empirical evidence about how people learn, and assessment of outcomes in STEM classrooms all point to a need to improve teaching methods to enhance learning and student persistence. Classroom approaches that engage students in "active learning" improve retention of information and critical thinking skills, compared with a sole reliance on lecturing, and increase persistence of students in STEM majors. STEM faculty need to adopt teaching methods supported by evidence derived from experimental learning research as well as from learning assessment in STEM courses. Evidence-based teaching methods have proven effective with a wide range of class sizes and increase learning outcomes even as enhancements of traditional lectures.

A significant barrier to broad implementation of evidence-based teaching approaches is that most faculty lack experience using these methods and are unfamiliar with the vast body of research indicating their impact on learning. The Federal Government could have a major impact by providing substantial support for programs that provide training for current and future faculty in evidence-based teaching methods and provide materials to support the application of such methods. Established programs run by the National Academies and the American Association of Physics Teachers/American Physical Society/American Astronomical Society have trained many faculty, and evaluations of these programs have demonstrated that they change the participants'teaching methods and have positive effects on student achievement and engagement. These programs provide successful models for replication and expansion.

Although evidence-based teaching methods do not necessarily require more resources than traditional lectures, the transition requires time and effort that can be costly for colleges and universities.

Given the Federal Government's interest in maintaining a strong STEM workforce, Federal support, in partnership with private and academic institutional investment, will be needed to initiate these changes, after which they can be sustained over the long term without external assistance.

Ongoing change toward the goal described here requires the ability to measure progress.

Metrics for excellence in undergraduate STEM education would provide tools for institutions, departments, funding agencies, external evaluators, accreditation agencies, students choosing where to study STEM subjects, and those designing innovative programs. Flexible criteria are needed to account for the wide range of institutions and disciplines that will use these tools to direct change.

Actions to Achieve Recommendation 1

1-1 Establish discipline-focused programs funded by Federal research agencies, academic institutions, disciplinary societies, and foundations to train current and future faculty in evidence-based teaching practices.

Successful programs should be expanded to reach 10% to 20% of the nation's 230,000 STEM faculty over the next five years. The expansion should make training available to faculty from diverse backgrounds to provide role models for all students and from all disciplines and types of institutions. Based on data from existing teaching training programs, it is reasonable to expect trained faculty to influence the teaching of 10 colleagues, making it possible to reach a substantial proportion of the STEM faculty through programs targeted at a subset of faculty.

Moreover, approximately 10% of the STEM faculty teach the introductory courses to first- and second-year college students. Therefore, the goal of reaching 10% to 20% of the STEM faculty directly could result in training most of those who teach in the first two years of college.

A total of $10-15 million per year over 5 years will be required for the training of 23,000 to 46,000 STEM faculty. Funds for this training should be derived from a combination of Federal programs academic institutions, disciplinary societies, and foundations.

To train future faculty, Federal research agencies should require all graduate students and postdoctoral fellows supported by federal training grants to receive instruction in modern teaching methods. A combination of training grant and institutional funds should be dedicated to this training effort.

1-2 Create a "STEM Institutional Transformation Awards" competitive grants program at NSF.

A competitive grants program should be designed to provide incentives for and facilitate teaching innovations at 2- and 4-year institutions. Grants should support model programs and electronic dissemination of successful practices. The grants program should have funding of $20 million per year, to support approximately 100 multi-year projects with average total support of $1 million over a 5-year period.

Funding could come from enactment of NSF's proposed Widening Implementation and Demonstration of Evidence-Based Reforms (WIDER) program at the Presidents' Fiscal Year 2012 requested level of $20 million annually.

1-3 Request that the National Academies develop metrics to evaluate STEM education.

To evaluate progress toward the goals presented in this report, campuses, funders, students, and accreditation agencies need a meaningful set of criteria by which to measure excellence in STEM education. NSF and the U.S. Department of Education should request The National Academies to lead an effort to develop metrics supported by empirical evidence that encourage and assess faculty practices and student learning.

Recommendation 2

Advocate and Provide Support for Replacing Standard Laboratory Courses with Discovery-Based Research Courses

Traditional introductory laboratory courses generally do not capture the creativity of STEM disciplines. They often involve repeating classical experiments to reproduce known results, rather than engaging students in experiments with the possibility of true discovery. Students may infer from such courses that STEM fields involve repeating what is known to have worked in the past rather than exploring the unknown. Engineering curricula in the first two years have long made use of design courses that engage student creativity. Recently, research courses in STEM subjects have been implemented at diverse institutions, including universities with large introductory course enrollments. These courses make individual ownership of projects and discovery feasible in a classroom setting, engaging students in authentic STEM experiences and enhancing learning and, therefore, they provide models for what should be more widely implemented.

Actions to Achieve Recommendation 2

2-1 Expand the use of scientific research and engineering design courses in the first two years through an NSF program.

The National Science Foundation should provide initial funding to replicate and scale-up model research or design courses, possibly through the existing Transforming Undergraduate Education in STEM (TUES) program or the Science, Technology, Engineering, and Mathematics Talent Expansion Program (STEP).

On the order of 30% of the existing programs across STEM disciplines could be focused on funding implemention of research courses at postsecondary academic institutions at an annual cost of approximately $12.5 million dollars (based on Fiscal Year 2010 funding levels). Based on the range of funding for Type 3 TUES grants and Type 1 STEP grants, about 10 proposals per year at an average level of $1.2 million could be awarded, in order to impact 100 campuses over the next 10 years.

Colleges and universities should seek to match NSF funding with private and philanthropic sources.

Research courses should be an encouraged element of STEM Institutional Transformation Awards. Because research courses will replace expensive introductory laboratory courses, they should not require ongoing external support once the transition is accomplished.

2-2 Expand opportunities for student research and design in faculty research laboratories by reducing restrictions on Federal research funds and redefining a Department of Education program.

Independent research on faculty projects is a direct way for students to experience real discovery and innovation and to be inspired by STEM subjects. All relevant Federal agencies should examine their programs which support undergraduate research and where there exists prohibitions, either in policy or practice, which would interfere with the recommendations of this report to support early engagement of students in research, these should be changed. Federal agencies should encourage projects that establish collaborations between research universities and community colleges or other institutions that do not have research programs. Cross-institutional research opportunities could be funded through redefinition of the Department of Education's $1 billion Carl D. Perkins Career and Technical Education program and by sharpening the focus of Federal investments in minority institutions.

Recommendation 3

Launch a National Experiment in Postsecondary Mathematics Education to Address the Mathematics-Preparation Gap

College-level skills in mathematics and, increasingly, computation are a gateway to other STEM fields. Today many students entering college lack these skills and need to learn them if they are to pursue STEM majors. In addition, employers in the private sector, government, and military frequently cite that they cannot find enough employees with needed levels of mathematics skills. This lack of preparation imposes a large burden on higher education and employers. Higher education alone spends at least $2 billion per year on developmental education to compensate for deficiencies. Also, introductory mathematics courses often leave students with the impression that all STEM fields are dull and unimaginative, which has particularly harmful effects for students who later become K-12 teachers. Reducing or eliminating the mathematics-preparation gap is one of the most urgent challenges—and promising opportunities—in preparing the workforce of the 21st century.

Closing this gap will require coordinated action on many fronts starting in the earliest grades. PCAST's earlier report on K-12 STEM education, *Prepare and Inspire: K-12 Education in Science, Technology, Engineering, and Math (STEM) for America's Future*, contains several recommendations that involve colleges and universities in this effort. In particular, it calls for the Federal Government to establish the objective of recruiting, preparing, and providing induction support for at least 100,000 new STEM middle and high school teachers who have majors in STEM fields and strong content-specific pedagogical preparation. This Administration has embraced this goal, and production of 1 million additional STEM graduates over the next decade could contribute substantially to meeting it.

The Federal Government has a critical role in supporting the development of a knowledge base to close the mathematics-preparation gap. For example, research into the best ways to teach math to older students so they can pursue STEM subjects in the first two years of college is badly needed. Some developmental mathematics courses have demonstrated effectiveness in increasing math proficiency among those not ready for college-level math and even in encouraging students intending to major in STEM subjects to persist to graduation and a STEM degree. Mathematics education research should explore the attributes of these successful classes and ways to disseminate best practices.

In the *Prepare and Inspire* report, PCAST also called for the creation of a mission-driven, Advanced Research Projects Agency for Education (ARPA-Ed) that would propel and support (1) the development of innovative technologies and technology platforms for learning, teaching, and assessment across all subjects and ages, and (2) the development of effective, integrated, whole-course materials for STEM education. Many of these advances would benefit not only K-12 education but also the developmental courses that many students need to pursue STEM fields during the first two years of college.

Actions to Achieve Recommendation 3

3-1 Support a national experiment in mathematics undergraduate education at NSF, the Department of Labor, and the Department of Education.

The National Science Foundation and the Departments of Labor and Education should support a multi-campus 5-year initiative aimed at developing new approaches to remove or reduce the mathematics bottleneck that is currently keeping many students from pursuing STEM majors. This national experiment should fund a variety of approaches, including (1) summer and other bridge programs for high school students entering college; (2) remedial courses for students in college, including approaches that rely on computer technology; (3) college mathematics teaching and curricula developed and taught by faculty from mathematics-intensive disciplines other than mathematics, including physics, engineering, and computer science; and (4) a new pipeline for producing K-12 mathematics teachers from undergraduate and graduate programs in mathematics-intensive fields other than mathematics. Diverse institutions should be included in the experiment to assess the impact of the intervention on various types of students and schools. Outcome evaluations should be designed as a collective effort by the participating campuses and funding agencies.

Approximately 200 experiments at an average level of $500,000 should be funded at institutions across the county, at an annual cost of $20 million per year for 5 years. As mathematics preparation issues vary across the postsecondary spectrum, a variety of sources will be needed to fund experiments at diverse institution types. Funds for these experiments could be derived from a combination of the Department of Education's proposed First in the World Initiative, possibly the Department of Labor's Career Pathways Innovation Fund or Trade Adjustment Assistance Community College and Career Training initiative, and a strategic focus on mathematics of NSF's Transforming Undergraduate Education in STEM (TUES) program or Science, Technology, Engineering, and Mathematics Talent Expansion Program (STEP) for the next 5 years.

Recommendation 4

Encourage Partnerships among Stakeholders to Diversify Pathways to STEM Careers

To take advantage of the breadth of available talent, non-traditional students should receive special attention. Adult and working students and those from backgrounds atypical of traditional STEM students may need alternative pathways to be successful in STEM disciplines. The concept of a "pipeline" to STEM competency and accomplishment needs to be superseded by the image of multiple pathways to these goals. All colleges and universities, including 2- and 4-year institutions, need better connections among themselves and with other institutions to provide more entry points and pathways to STEM degrees.

Actions to Achieve Recommendation 4

Establishing and supporting pathways will require a coordinated effort among diverse institutions. The Federal Government can lead this effort and encourage the necessary partnerships through strategic planning, reallocation of funds, and leadership.

4-1 Sponsor at the Department of Education summer STEM learning programs for high school students.

The Department of Education should roll-out the summer learning programs authorized in the 2007 America Competes Act (in an amendment introduced by then-Senator Obama) to provide mathematics instruction and hands-on STEM experiences for rising high school juniors and seniors. The programs should be funded by partnerships among the Federal Government, states, local entities, and private industry. Based on the size of National Science Foundation's former Young Scholars Program for summer institutes, we recommend an investment of $10 million to fund approximately 100 projects reaching on the order of 5000 students, annually, with significant cost sharing with academic institutions and private investors.

4-2 Encourage pathways from 2- to 4-year institutions through an NSF program and expanded definition of a Department of Labor Program.

The mission of the Department of Labor's Trade Adjustment Assistance Community College and Career Training initiative should be expanded beyond development of important partnerships between community and technical colleges and employers in the private sector to encourage scientific research and engineering design exchanges across two- and four-year institutions. Alternatively, these activities could be funded through a strategic focus of the Department of Labor's Career Pathways Innovation Fund on research partnerships. NSF's Advancing Technical Education program could also be focused on cross institutional collaborations. The bridges described here should provide authentic STEM experiences for community college students on the four-year campus and allow students to develop relations with faculty and the college or university community to ease the potential transition from a 2- to 4-year institution or to provide advanced experiences for students who do not pursue a four year degree.

4-3 Establish public-private partnerships to support successful STEM programs.

To enhance students'STEM readiness, the Federal Government should engage private industry and foundations to support successful programs that create bridges between high schools and colleges and between 2- and 4-year institutions and ensure that programs incorporate learning standards and content consistent with industry-recognized skills.

4-4 Improve data provided by the Department of Education and the Bureau of Labor Statistics to STEM students, parents, and the greater community on STEM disciplines and the labor market.

To promote pathways to STEM careers for non-traditional students, the Federal Government should provide current and comprehensive data on STEM jobs. Today, public and private employers of STEM professionals lack data about the skills, choices, and availability of STEM workers. To produce needed information, the 1988 cohort and the *High School and Beyond* cohort should be resurveyed; the Department of Education should devote more resources to tracking students from high school into their careers; and the Bureau of Labor Statistics should redefine employment categories to include in "STEM" the breadth of jobs that require STEM skills, such as medical careers and advanced manufacturing professions.

Recommendation 5

Create a Presidential Council on STEM Education with Leadership from the Academic and Business Communities to Provide Strategic Leadership for Transformative and Sustainable Change in STEM Undergraduate Education

The leadership of higher education and STEM-enabled businesses needs to be inspired to generate sweeping changes in higher education to produce the workforce America needs.

Toward this end, we recommend that the President, via Executive Order, form a Presidential Council on STEM Education to provide advice and leadership on postsecondary STEM education. The council should include members that represent the breadth of academic institutions, professional societies, businesses, and private foundations involved in the development and use of human capital in STEM fields. Based on the guidance provided in this report, the council should make recommendations that advance the quality of postsecondary STEM education through all mechanisms available to the President. The council could provide a forum for leaders in the public and private sectors to weigh in on the development and deployment of metrics to evaluate STEM departments (Recommendation 1) and to design collaborative coalitions to support initiatives in STEM education (Recommendation 4), including expanding internship programs in industry and connecting industrial research agendas with research courses (Recommendation 2). In addition, it could provide advice and review for the National Experiment in Math Undergraduate Education (Recommendation 3) and could conduct further study of the math education issue, if necessary.

Overview of Pcast Recommendations to Engage and Excel in Undergraduate Science, Technology, Engineering, and Mathematics (Stem) Education

Recommendation 1: Catalyze widespread adoption of empirically validated teaching practices.

1-1 Establish discipline-focused programs funded by Federal research agencies, academic institutions, disciplinary societies, and foundations to train current and future faculty in evidence-based teaching practices.

1-2 Create the "STEM Institutional Transformation Awards" competitive grants program at NSF. 1-3 Request that the National Academies develop metrics to evaluate STEM education.

Recommendation 2: Advocate and provide support for replacing standard laboratory courses with discovery-based research courses.

2-1 Expand the use of scientific research and engineering design courses in the first two years of postsecondary education through an NSF program.

2-2 Expand opportunities for student research and design in faculty research laboratories by reducing restrictions on Federal research funds and redefining a Department of Education program.

Recommendation 3: Launch a national experiment in postsecondary mathematics education to address the mathematics-preparation gap.

3-1 Support a national experiment in mathematics undergraduate education at NSF, the Department of Labor, and the Department of Education.

Recommendation 4: Encourage partnerships among stakeholders to diversify pathways to STEM careers.

4-1 Sponsor at the Department of Education summer STEM learning programs for high school students.

4-2 Expand the scope of a Department of Labor Program and focus an NSF program to encourage pathways from 2-to 4-year institutions.

4-3 Establish public-private partnerships to support successful STEM programs.

4-4 Improve data provided by the Department of Education and the Bureau of Labor Statistics to STEM students, parents, and the greater community on STEM disciplines and the labor market.

Recommendation 5: Create a Presidential Council on STEM Education with leadership from the academic and business communities to provide strategic leadership for transformative and sustainable change in STEM undergraduate education.

Working Group Report

PCAST STEM Undergraduate Education Working Group

Co-Chairs
S. James Gates, Jr.*
John S. Toll Professor of Physics
Director, Center for String and Particle Theory
University of Maryland, College Park

Jo Handelsman
Howard Hughes Medical Institute Professor and
Frederick Phineas Rose Professor, Molecular,
Cellular and Developmental Biology
Yale University

G. Peter Lepage
Professor of Physics
Harold Tanner Dean, College of Arts and Sciences
Cornell University

Chad Mirkin*
Rathmann Professor, Chemistry,
Materials Science and Engineering,
Chemical and Biological Engineering
and Medicine Director,
International Institute for Nanotechnology
Northwestern University

Working Group Members

Joseph G. Altonji
Thomas DeWitt Cuyler Professor
of Economics Yale University

Peter Bruns
Vice President for Grants and
Special Programs (Retired)
Howard Hughes Medical Institute (HHMI)

Carol Christ
President Smith College

Isaac Crumbly
Associate Vice President for Career and
Collaborative Programs
Founder,
Cooperative Development Energy Program
Fort Valley State University

Emily DeRocco
President
The Manufacturing Institute
Senior Vice President
National Association of Manufacturers

Brian K. Fitzgerald
Chief Executive Officer
Business Higher Education Forum

Mark Gorenberg*
Managing Director
Hummer Winblad Venture Partners

Shirley Ann Jackson*
President
Rensselaer Polytechnic Institute

Neal Lane
Senior Fellow in Science and Technology Policy
James Baker Institute for Public Policy
Rice University

Tom Luce
Chief Executive Officer
National Math and Science Initiative

Judy Miner President
Foothill College

Suzanne Ortega
Senior Vice President for Academic Affairs
University of North Carolina

Ed Penhoet*
Director, Alta Partners
Professor Emeritus, Biochemistry
and Public Health
University of California, Berkeley

Calvin Phelps
National Chair
National Society of
Black Engineers (NSBE)

Staff

Danielle Evers
AAAS Science and Technology
Policy Fellow

Daniel Schrag†
Sturgis Hooper Professor of Geology
Professor, Environmental Science and Engineering
Director, Harvard University Center for Environment
Harvard University

Dan Schwartz
Professor
Stanford University School of Education

Candace Thille
Director of the Open Learning Initiative Carnegie
Mellon University

Deborah Stine
Executive Director

Writers

Steve Olson
Science Writer

Donna Gerardi Riordan
Science Writer and Policy Analyst

I. Introduction

Importance of STEM

Throughout the 20th century, science, technology, and higher education were drivers of innovation in the U.S. economy. The rapid expansion of the research and development enterprise after World War II—which was enabled by the growth of higher education and corresponding increases in the number of college graduates with expertise in science, technology, engineering, and mathematics (STEM)—led to strong economic performance, good jobs, and thriving new industries driven by new technologies.

The United States is now putting its future at risk by forfeiting its historical strengths in STEM education. The proportion of STEM degrees among all college graduates has been falling for the past decade.[1] Without action, it is likely that this proportion will continue to drop as groups that have historically earned fewer STEM degrees on average than white men become a larger majority of college students.[2]

As has occurred previously —with the 1862 Federal support for the establishment of land grant colleges, for example, and almost a century later with the response to the launch of

† PCAST member.

Sputnik—the Nation has reached a decision point. The United States could renew its commitment to education—and especially STEM education—or it could risk creating a permanent economic gap among American workers at a time of dramatic demographic transition and enhanced global economic competition.

The need for STEM knowledge extends to all Americans. The products of science, technology, engineering, and mathematics play a substantial and growing role in the lives of all Americans. A democratic society in which large numbers of people are unfamiliar or uncomfortable with scientific and technological advances faces a great economic disadvantage in globalized competition. Achieving scientific and technological literacy among our citizenry is a complex topic that differs in important ways from the challenge of training STEM professionals and is beyond the scope of this report; we hope that this topic will become the focus of future study. Nevertheless, the actions we recommend, though not specifically targeted at achieving broad STEM literacy, will affect STEM literacy among the college-educated citizenry.

One Million Additional College Graduates with STEM Degrees

Several analyses point to the need to add to the American workforce over the next decade approximately 1 million more STEM professionals than the U.S. will produce at current rates.[3,4,5] The exact projections vary somewhat depending on the job definitions and assumptions embodied in the models, but the studies produce results on the same order of magnitude. For example, one analysis by the Center on Education and the Workforce at Georgetown University shows that between 2008 and 2018, STEM occupations will increase from 5.0% of the jobs in the U.S to 5.3%, an increase that is equivalent to 1 million jobs[6] and that 92% of STEM jobs by 2018 will require at least some postsecondary education and training.[7] This projection also aligns with the President's goal of the United States regaining its lead in the world in the number of young people graduating from college. If the STEM fields are going to take part in this growth of college-educated Americans, the number of STEM degrees earned must increase by about 1 million over the next decade.

In the past, the United States has relied on foreign-born STEM professionals to satisfy unmet workforce demands, and these employees have made important contributions to the U.S. economy. But the U.S. is not guaranteed a continuing future supply of international workers in STEM fields because education and employment opportunities are increasing in numbers elsewhere. Reliance on foreign nationals makes our security and economy vulnerable as their home countries become more attractive and their STEM-trained workers return from the U.S. to serve the needs of their homelands. Moreover, STEM-related jobs are among the best our economy offers, as evidenced by their high wages and lower unemployment rates than in other sectors.[8,9,10] The increased supply of jobs in these fields will offer an opportunity to reduce income inequality in the United States.[11,12] This opportunity can be captured only by increasing the number of U.S.-born college graduates with training in STEM fields from all demographic sectors of U.S. society.

The U.S. currently graduates about 300,000 bachelor and associate degrees in STEM fields annually;[13] thus, between 2012 and 2022, the U.S. can be expected to produce approximately 3 million STEM degrees. To meet the goal of an additional 1 million STEM college graduates in the next decade, we would need to graduate an additional 100,000 per year, representing an approximately 33% increase over current production rates.

This goal is justified and will be feasible with strategic actions. (See Appendix D for a more extensive analysis of the need for 1 million additional STEM workers.)

Beyond STEM Professionals

In this report, STEM professionals are defined as those with degrees in STEM areas who are trained and work as STEM practitioners.

In addition to the need for more STEM professionals, there is also a national need for more workers with some STEM training. These "STEM-capable" workers are able to use knowledge and skills from STEM fields but work in areas that are traditionally considered non-STEM fields. The ranks of the STEM-capable workforce are expanding as this skill set comes to represent an increasingly valued commodity in many fields.

For example, physicians, nurses, and other health workers and advanced manufacturing professionals generally are not categorized as "STEM professionals," yet many of these jobs draw heavily on STEM knowledge and skills, and represent some of the most rapidly growing or wealth producing sectors of the U.S. economy.

Another group that is not counted in economic projections as STEM professionals are K-12 teachers with strong STEM skills, whose shortage has become a national crisis. (See Appendix D for further description of STEM skills categories).

Since none of these substantial groups is counted among the needed STEM professionals in the economic projections we cite here, the size of the future workforce needing STEM training may substantially exceed the addition of the estimated 1 million STEM professionals.

The recommendations we present should affect the college-educated population generally by increasing interest in and knowledge of STEM subjects among graduates of diverse fields, thereby broadening the impact of the recommended actions beyond STEM professionals.

Engage to Excel

The themes guiding this report have broad application to leaders, faculty, and students in academia, industry, and government.

The title of this report, *"Engage to Excel,"* applies to individuals across these groups. Students must be engaged to excel in STEM fields.

To excel as teachers, faculty must engage in methods of teaching grounded in research about why students excel and persist in college. Moreover, success depends on the engagement by great leadership.

Leaders, including the President of the United States, college, university and business leadership, and others, must encourage and support the creation of well-aligned incentives for transforming and sustaining STEM learning.

They also must encourage and support the establishment of broad-based reliable metrics to measure outcomes in an ongoing cycle of improvement.

II. STRATEGIES: THE FIRST TWO YEARS

How to Fill the Need?

In the United States, fewer than 40% of the students who enter college with the intention of majoring in a STEM field complete a STEM degree. Most of the students who leave STEM fields switch to non-STEM majors after taking introductory science, math, and engineering courses.[14] Many of the students who leave STEM majors are capable of the work, making the retention of students who express initial interest in STEM subjects an excellent group from which to draw some of the additional one million STEM graduates. Research on the exodus from STEM disciplines shows that many students who transfer out of STEM majors perform well, but they describe the teaching methods and atmosphere in introductory STEM classes as ineffective and uninspiring.[15,16]

Others do not perform well despite interest and aptitude and would benefit from alternative teaching methods, tutoring, or other experiences demonstrated to improve performance in STEM subjects. Merely increasing retention from 40% to 50% would translate to an additional 72,500 STEM degrees per year, comprising almost three-quarters of the 1 million additional STEM graduates needed over the next decade.

Although women and members of minority groups now constitute approximately 70% of college students, they are underrepresented among students receiving undergraduate degrees in STEM subjects (approximately 45 percent). These students are an "underrepresented majority" that must be part of the route to excellence.[17] Members of this group leave STEM majors at higher rates than others and offer an expanding pool of untapped talent.

Some campuses have shown that differences in performance and retention between traditional STEM majors and members of the underrepresented majority can be reduced substantially by several simple changes in campus or classroom practices (e.g., see Appendices F and G).[18,19,20,21] The underrepresented majority is a large underutilized source of potential STEM professionals and deserves special attention.

The current system of STEM education has effectively trained many STEM workers, including most of the current STEM workforce. However, its longevity is not evidence that it cannot be improved or that this system will be successful with today's student body. Indeed, extensive evidence points to a need to do better. STEM disciplines have substantially lower rates of retention than do the social sciences and humanities. [22,23,24]

Furthermore, many of those who leave STEM majors express dissatisfaction with the teaching of STEM classes.[25,26] This should be seen as a national crisis of STEM teaching, yet many STEM faculty members believe that this "weeding out" process is in the best interest of their disciplines and the larger national interest.

If many of those who leave performed well in introductory STEM courses, and many others could be helped to succeed, then it is unreasonable to conclude that this attrition represents an effective selection process that is maximally beneficial to STEM fields.[27]

The first two years of college are the most critical to retention and recruitment of STEM majors. The STEM courses in these years are also a shared feature of all types of 2- and 4-year colleges and universities—community colleges, comprehensive universities, liberal arts colleges, research universities, and minority-serving institutions. In addition, STEM courses in the first two years are all the STEM courses that most future K-12 teachers are going to experience in college.

The amount they learn, the models of STEM teaching, and their attitudes towards STEM disciplines will have an enormous impact on their future teaching. For all these reasons, a focus on improving STEM courses taken early in college offers potentially enormous benefits to STEM fields. Therefore, this report focuses on actions that will influence the quality of STEM education in the first two years of college.

Persistence of Students in STEM Majors

Research indicates that student persistence in a STEM degree is associated primarily with three aspects of their experience. The first concerns *intellectual engagement and achievement.* Compared with students in traditional lectures, students who play an active role in the pursuit of scientific knowledge learn more and develop more confidence in their abilities, thereby increasing their persistence in STEM majors. This engagement can be accomplished in both the classroom and research lab. Many types of classroom instruction that engage students in thinking or problem-solving increase learning and enhance attitudes toward STEM fields. These gains translate into better retention of students in STEM majors. For example, students in traditional lecture courses were twice as likely to leave engineering and three times as likely to drop out of college entirely compared with students taught using techniques that engaged them actively in class.[28] In a randomized trial at the University of Michigan, students who engaged in sophomore research with a professor were much less likely to leave STEM majors than those who did not. The effects were observed among all groups, including white, African American, and Hispanic students.[29]

The second aspect of a student's experience that affects persistence is *motivation.* Motivation is partially intrinsic but also is modulated by the college environment. A key in maintaining student motivation is having role models. The majority of U.S. STEM faculty are white, male, able-bodied, and middle class and have had many role models with whom to identify. Role models who are women and ethnic minorities increase the performance and retention of students in those same groups.[30,31,32] Financial concerns,[33] lack of encouragement from family members,[34] and a deficit of peers from similar backgrounds can erode self-confidence and the will to remain in STEM majors.[35]

A student's belief about barriers or pathways to success in an academic field also influences motivation. For example, students who believe that hard work is the key element in success are more likely to interpret negative feedback as guidance for improvement, whereas students who believe that intrinsic ability determines a person's success are more likely to take negative feedback as a negative assessment of their ability to perform.[36] Courses that have very low average grades on exams can differentially discourage the latter group of students from continuing in STEM majors.[37]

Some simple experiences have been shown to have large effects on performance and persistence. In one study, female subjects instructed to focus on the similarities between men and women performed better on a math exam and expressed less preference for typically feminine careers than students who received instructions that were not directed at gender.[38] A dramatic effect was achieved in performance on physics exams by having students write for 15 minutes about their values.

This exercise only affected the women, thereby closing a rather substantial achievement gap between men and women in the class.[39] A recent paper reported a study of students who experienced a one-time intervention in which they were asked to read a short article about adversity in college and then write an essay and speak about it. Over a 3-year period, the African American students who experienced the intervention had grade point averages a full grade higher than those who did not experience the session, had fewer health problems, and had a greater sense of well-being than African American students in the control group.[40] In addition, many studies show that participation in research improves motivation and active participation in subsequent courses for all students.[41]

The third aspect of a student's experience that affects persistence is *identification with a STEM field*. Recent work suggests that identification with a group or community of STEM professionals may overshadow many other factors in determining persistence.[42,43,44]

Developing meaningful relationships with peers and instructors, involvement in study groups, and participation in a research laboratory all are associated with reduced departures from STEM fields.[45,46,47,48,49,50,51,52,53,54] Interventions to enhance minority students' identification with a field equalize retention between minority and majority students,[55] indicating the need for more focused programs that emphasize student engagement as part of a STEM community.

Strategies to Achieve Engagement and Excellence in STEM Learning

All three of the aspects of student experience discussed above must be addressed to increase retention among STEM students. The key strategies that we propose in this report fall into three broad categories:

1. Adopt STEM teaching strategies that emphasize student engagement. The lecture has been a mainstay of higher education since the word "lecture" was created in the 14th century, and today most introductory STEM courses are taught largely through lectures. Extensive research on how the human brain learns indicates that diversifying teaching methods enhances critical thinking skills, long-term retention of information, and student retention in STEM majors.[56,57,58,59,60,61,62,63,64] Moreover, these active learning techniques benefit *all* students and can close the achievement gap between ethnic groups and men and women. We therefore recommend that STEM faculty learn how to use and incorporate highly effective teaching methods into their introductory STEM courses, including the opportunity to generate knowledge through research. These methods should include research courses, other forms of active student engagement, and learning assessment as part of a continued cycle of improvement in STEM education.

2. Provide all students with the tools to excel. Many students arrive in college without sufficient study skills, math proficiency, or identification as a scientist, engineer, or mathematician. These three contributors to success in STEM disciplines also are distributed differentially among ethnic and socioeconomic groups as well as between men and women. These are key foci for change that will reduce the achievement gap and increase retention of students in STEM courses. We therefore recommend high school to college bridge programs and other mechanisms to improve study skills, identification with STEM fields, and

particularly math preparation. The POSSE foundation[65] provides a model from which key features can be used as a gold standard for bridge programs:

- A rigorous selection process for students with academic excellence, leadership potential, and interest in STEM fields.
- Enrichment programs and cohort events to build community and a support network for students
- Academic programs during the summer after high school to enable college readiness
- Mentoring, advising, and tutoring at college, including assistance finding a research laboratory

3. Diversify pathways to STEM degrees. There was a time when most people who attended college were single white men, had high school diplomas, started college at age 18, graduated in 4 years, had all the academic preparation needed to succeed, and had few family responsibilities. In the 21st century, this is not true. Today, students come from diverse backgrounds, have widely divergent levels of preparation, may be returning to college after years in the workforce or serving in the U.S. military, and often are employed while in college to support themselves and families. Higher education needs to acknowledge these differences among students and work to accommodate them by creating more entry points and pathways to STEM degrees. At the beginning of the 21st Century, the concept of a "pipeline" to STEM competency and accomplishment needs to be replaced by a system of multiple pathways to these goals.

III. Barriers and Challenges

Institutional and Individual Barriers Demand a Multifaceted Approach to Catalyzing Change

The strategies introduced in the previous section have the potential to transform undergraduate STEM education, but change in academia is slow and hard. The status quo is favored for many reasons, such as existing incentive structures and traditional practices. In this chapter of the report, we describe the most significant barriers to implementing the three strategies to achieve excellence and engagement. The next chapter presents a multifaceted set of actions that the Federal Government can take to encourage change and reduce or circumvent the barriers.

Faculty Lack Knowledge of Evidence-Based Teaching

Despite what is a now vast body of research about how people learn and which teaching methods are most effective at transmitting knowledge and building critical thinking skills, most STEM faculty members have neither the time nor the incentives to find, read, and evaluate the literature or the teaching methods derived from it. Most teach using methods by which they were taught. Access to and implementation of modern assessment of learning is similarly distant from and inaccessible to the typical faculty member. Few opportunities for formal training in STEM teaching and assessment exist, and those that do are hard to find.

Lack of Facilitation and Rewards for Good Teaching

Instituting more effective institutional approaches than lecturing will require convincing a large segment of those teaching STEM courses that they can teach more effectively while still meeting all of their professional obligations (including teaching multiple courses, conducting research, and serving their institutions and disciplines).

Today, faculty members still face several major obstacles to changing their teaching practices:

- Insufficient time to acquire the latest information on the most effective evidence-based teaching practices,
- lack of individual rewards for teaching, even at liberal arts colleges, where salaries and advancement more closely correlate with publication rate than teaching quality,
- lack of departmental rewards and expectations for good teaching.

The current incentive system for most STEM faculty is focused on research and not teaching. It therefore discourages the expenditures of time and effort required to surmount the obstacles cited above. As things stand, it seems untenable to expect faculty to become proficient practitioners of a research field as well as experts on the literature on effective evidence-based teaching practices. They need to be provided with tools and information that they can readily use in their teaching. To increase recognition of the importance of teaching in research institutions, it will be critical to have leadership from presidents and provosts to galvanize faculty through resources and rewards. Department chairs are critical to that effort because they have the most direct impact on teaching in STEM departments. Some of the changes are easy and inexpensive. For example, a department's web-page might simply provide a set of learning goals for students in their major; the process of agreeing on these learning goals would immediately elevate the visibility and importance of teaching and likely improve it as well. A department's website might also list faculty members who are outstanding teachers and provide evidence for their excellence. Some changes will be harder. Adding a requirement for teaching excellence to tenure guidelines has been accomplished at many research universities,[66,67] but will be highly controversial at others. Some changes will require new resources. Revamping courses and curricula is difficult and requires time that must be subsidized. Resources can be used to influence how faculty spend their time and will be essential to seed transformation and institutionalization of improved STEM teaching.

Limited Resources

Most universities have felt the economic realities of the last few years. Some struggle to provide the most basic elements of their curricula, so the idea of putting time and resources into new teaching approaches and programs may seem unrealistic. Leadership will need to address this issue through reallocation of existing resources, strategic fundraising, and securing financial assistance from private funders and State and Federal grants. However, the strategies proposed here require little expansion of the introductory courses. Increasing retention of students *beyond* the introductory courses will generate most of the new STEM majors.

Grading and Workload across Majors

Some students avoid or abandon STEM majors because they believe that their GPAs are likely to be lower in STEM courses than in humanities, business and management, or social sciences and that the workload is greater. They are correct at most universities. However, faculty can make it known that they are available to help students learn to ensure that they do as well as possible in their courses. STEM faculty also can make their courses so engaging that students will be inspired by STEM fields and persist in STEM majors despite the workload. Most students who intend to major in a STEM field have an intrinsic interest in STEM subjects that can compensate for the differences between STEM and other courses. Arbitrary depression of grading scales in STEM courses should be discontinued. These practices artificially reward students for majoring in non-STEM disciplines, especially for students who feel pressure from financial aid, GPA requirements, or graduate school admissions.

Institutional Isolation

Many two-year and non-research institutions do not have the programs or resources to offer students a full suite of research opportunities. Addressing this will require collaborations among academic institutions as well as between academia and industry to expand the opportunities for students beyond their own institution.

Challenge of Change

People are usually resistant to change. One reason that many faculty may maintain traditional teaching practices is that they have been successful in their fields and therefore assume that the educational approaches that taught them so effectively are appropriate for all students. But resistance to change is human and has been confronted successfully in numerous other settings. The study of individual, organizational, and cultural change is a sophisticated field that can inform the design of transformation strategies for STEM education in the first two years of college.

The fact that lecturing remains the overwhelmingly predominant form of instruction at the postsecondary level when there are hundreds of papers showing better ways to teach indicates that more than inertia is at work. The incentives for both the academic department and the individual faculty member at research universities are focused on maximizing research success, and this system has worked extremely well to maintain a powerful research engine in higher education. However, there are few, if any, counter-balancing incentives linked to desired educational outcomes, and there are often disincentives. One that exerts an overwhelming influence on junior faculty is the current tenure decision system. Though increased attention is now being paid to teaching effectiveness, tenure decision processes still push mainly in the opposite direction. Even if junior faculty come to an institution with the passion and determination to achieve teaching excellence, they can easily feel, and are often advised by their more senior colleagues, that teaching innovations should wait until after they have achieved tenure.

Effective incentives require good metrics for measuring accomplishment—metrics by which departments and individual faculty members can be compared and held accountable. Although research will always be the hallmark of the research university and must be valued and rewarded, the ideal faculty incentive system is based on both teaching and research accomplishments. For the incentive system to be meaningful, metrics for teaching quality must be credible.

To achieve the goals presented in this report, colleges and universities need to change their institutional and reward structures. In the last few decades, some extraordinary, sweeping changes have been deliberately instigated and studied in other societal areas. For example, the nearly universal familiarity in the United States with the idea of a "designated driver," previously unknown in our society, was achieved in three years because of one person's vision and action. Such campaigns provide guidance for designing similarly transformative initiatives.

Table 1. Actions to induce cultural change

Create a sense of urgency	Reward change
• Identify credible guiding teams	• Ensure repeated exposure to message
• Create vision	• Provide checklists to measure progress
• Communicate vision and progress	• Create community for transformation leaders
• Facilitate change/remove obstacles	• Use diverse, concerted drivers to generate a tipping point
• Generate belief in successful movement	• Reward change

Sources:

Gladwell, M. (2002). The Tipping Point: How Little Things Can Make a Big Difference. New York, NY: Little, Brown & Co.;

Heath, C. and D. Heath (2010). Switch. New York, NY: Broadway Books.

Kotter, J. (1996). Leading Change. Boston, MA: Harvard Business Review Press.

Shapiro, A. (2003). Creating Contagious Commitment: Applying the Tipping Point to Organizational Change. Hillsborough, NC: Strategy Perspective.

Based on the theory and practice of cultural change, a number of steps must be accomplished to effect lasting change for STEM education (Table 1). Key elements to be addressed include human tendencies such as resistance to change, complacency, and cynicism; practical obstacles such as lack of resources and know-how; communication challenges including lack of awareness of the problem or successful solutions; and lack of reinforcements to foster change among individuals and institutions. The recommendations we make in the next chapter are focused on addressing the challenges, generating an environment, and establishing processes that will induce and sustain transformative change.

IV. A Multi-Facted Approach: Reaching the Tipping Point

No single strategy will generate 1 million additional undergraduate STEM degrees over the next decade, because the challenge has many dimensions. It entwines facts and logic with academic culture, incentives, and belief systems. Therefore the recommendations presented here address various stakeholders and use both tangible resources and persuasion to inspire and catalyze change in undergraduate STEM education. By attacking the issue from a number of angles with various tools, including public exhortation, faculty incentives, resources, information, and institutional connections, the concerted forces can reach a point at which the movement takes on a momentum of its own and leads to sweeping change.

Barriers to change vary with institution type and context. Some institutions may respond to a desire to be on the cutting edge of education, some to new resources, and others to the desire to maintain funding for and prestige of their graduate programs. Some faculty will be interested in change but will not know how to accomplish it; others will be waiting to hear from their administrations that this change is important and will be rewarded. Some students might benefit most from engaging in research, while others might be more in need of bolstering their math skills. Therefore, we propose promoting change with actions that address diverse students, faculty, departments, institutional leadership, industrial interests, and professional societies.[68] Our recommendations aim to overcome many barriers, from lack of faculty time for studying the education literature to the inability of students to re-enter college after they take a break from their education.

A number of steps must be accomplished to effect lasting change. Needed elements include a combination of rational thinking, a sense of urgency, community facilitation, cooperative action among key players, individual and group rewards, and visible success stories.

When the point is reached where ongoing change no longer depends on interventions by the Federal Government, the importance of engagement and excellence in STEM education will be part of the academic lexicon on every institution's agenda, and will be widely accepted as beneficial to students, faculty, and society. When this point is reached, resources for the recommendations below will be incorporated into the base budgets of many institutions, graduate students will not remember a time when science was taught by lectures alone, and having metrics to evaluate excellence in STEM education will be routine.

Recommendations

The President's Council of Advisors on Science and Technology (PCAST) proposes five overarching recommendations to transform undergraduate STEM education during the transition from high school to college, and during the first two years of undergraduate STEM education:

1) Catalyze widespread adoption of empirically validated teaching practices.
2) Advocate and provide support for replacing standard laboratory courses with discovery-based research courses.

3) Launch a national experiment in postsecondary mathematics education to address the mathematics preparation gap.
4) Encourage partnerships among stakeholders to diversify pathways to STEM careers.
5) Create a Presidential Council on STEM Education with leadership from the academic and business communities to provide strategic leadership for transformative and sustainable change in STEM undergraduate education.

Each of these recommendations will be explained in more detail below.

Recommendation I

Catalyze Widespread Adoption of Empirically Validated Teaching Practices

Rationale for Recommendation 1

Evidence-Based Teaching

Thinking like a STEM professional requires acquisition of information, habits of mind, skills, and an identity embedded in a STEM discipline.[69] Such diverse attributes are unlikely to be learned most effectively through one mode of teaching.

Yet most introductory STEM courses taken in the first two years of college are dominated by lectures and multiple choice tests. A substantial empirical literature has demonstrated that alternative models of instruction can achieve many important learning outcomes more effectively than current practice (Table 2). (For a discussion of the learning literature, see Appendix F.) STEM educators can take a more scientific approach to teaching by basing classroom choices on research evidence rather than habits and traditions.

Table 2. Types of active learning that have been demonstrated to enhance learning

Types of active learning with feedback	Examples of studies that demonstrate enhanced learning
Small group discussion and peer instruction	Anderson et al. (2005); Armbruster et al. (2009); Armstrong et al. (2007); Beichner et al. (1999); Born et al. (2002); Crouch and Mazur (2001); Fagen (2002); Lasry et al. (2008); Lewis and Lewis (2005); McDaniel (2007a, 2007b); Rivard and Straw (2000); Tessier (2004 and 2007); Tien et al. (2002)
Testing	Steele (2003)
One-minute papers	Almer et al. (1998); Chizmar and Ostrosky (1998); Rivard and Straw (2000)
Clickers	Smith et al. (2009, 2011)
Problem-based learning	Capon and Kuhn (2004); Preszler et al. (2007)
Case Studies	Preszler (2009)
Analytical challenge before lecture	Schwartz and Bransford (1998)
Group tests	Cortright et al. (2003); Klappa (2009)
Problem sets in groups	Cortright et al. (2005)
Concept mapping	Foncesca et al. (2004); Prezler (2004); Yarden et al. (2004)
Writing with peer review	Pelaez (2002)
Computer simulations and games	Harris et al. (2009); McDaniel et al. (2007); Traver et al. (2001)
Combination of active learning methods	Freeman et al. (2007); O'Sullivan and Cooper (2003)

"Evidence-based" teaching, also known as "scientific teaching",[70] has two features. First, it involves choosing teaching methods based on research about how people learn and on proven teaching methods. Second, it involves using assessment of learning to determine whether students are meeting stated learning goals. Generally, approaches that most effectively transmit information and build critical thinking skills require that students are actively engaged in the process of and receive feedback while learning.[71]

Classroom approaches that engage students actively have been shown to increase retention of information, build critical thinking skills, induce more positive attitudes toward STEM disciplines, and increase retention of students in STEM majors.[72,73] Diverse methods that engage students in "active learning" have been successfully implemented in a large range of classroom sizes and can be done with an *increase,* not a decrease, in coverage of content. Most surprisingly to many instructors is the increase in retention of information, deep understanding, and student attendance and enthusiasm in class that result from a diversification of teaching approaches beyond lectures (see Table 2 for references).

Three types of research studies demonstrate the effects of evidence-based teaching methods on learning and retention in STEM degrees.

The field of cognitive psychology has constructed a substantial literature of randomized trials with carefully controlled variables in which different types of teaching are used and learning is measured with a single instrument. These studies consistently show that learning and retention of knowledge as well as acquisition of higher order thinking and reasoning skills are better with many types of active learning than lectures alone.[74,75]

- Designed experiments that compare learning between two STEM courses in which the material is presented with active or passive (largely traditional) means show that the introduction of active learning into STEM courses of all sizes and disciplines induces more learning than lectures alone.[76,77,78] In some of these experiments the instructor is the same in both classes, and in some the instructors differ.
- Retrospective analyses of student performance in courses or curricula, which resemble epidemiological studies in human health, also demonstrate more learning with active than passive methods.[79]

While each of these lines of evidence has limitations, together they create a compelling body of research indicating that student learning can be enhanced by any of a large number of interventions that induce active engagement of students in the course material.

A more extensive discussion of this research, a summary of more than 100 papers in the field, and a discussion of the experimental approaches used to avoid some of the obvious pitfalls of this type of research are presented in Appendix F.[80]

Technology to Improve Learning

Technology can be used in far more meaningful ways than is currently typical in STEM classrooms. In addition to its use to save cost and time (e.g., putting a textbook, lecture, or assessment on-line) and disseminate learning and assessment materials (e.g., portals that enable educators to search for lessons online), far more dramatic change in education can be achieved with technologies that create a "cycle of innovation." Globally available and shared assessment tools can evaluate student learning and feed learning data into central databases for researchers as well as learners and teachers, leading to continuous improvement of teaching and learning. As knowledge about learning evolves, this cycle of innovation can provide a natural route for continuous experimentation, with immediate feedback for many different types of classrooms and the provision of information to teachers about which methods are successful in particular settings. This process also can aid in providing researchers in the cognitive sciences with the data to develop generalizable principles about learning. Teachers will be able to adapt teaching methods to maximize learning based on both specific data about their current students and research conducted across many classrooms. Expansive use of this innovation model, facilitated through use of technology, will provide ongoing improvement based on evidence, much like the software industry provides product updates based on user reports and patterns of use.

Actions to Achieve Recommendation 1

1-1 Establish discipline-focused programs funded by Federal research agencies, academic institutions, disciplinary societies, and foundations to train current and future faculty in evidence-based teaching practices.

Federal agencies, in particular NSF, should fund expansion of existing programs designed to train current faculty in effective college teaching methods and provide materials to support the application of such methods. These efforts should be undertaken in partnership with disciplinary societies and foundations, and with matching funds for faculty participation contributed by academic institutions. The expansion should make training available to faculty from diverse backgrounds to provide role models[81] for all students and from all disciplines and types of institutions. Examples of model programs include the National Academies' Summer Institutes for Undergraduate Education in Biology[82] and the American Association of Physics Teachers (AAPT)/ American Physical Society (APS)/American Astronomical Society's (AAS) Physics and Astronomy New Faculty Workshop and the Association for Computing Machinery's Special Interest Group on Computer Science Education (SIGCSE) Symposium.[83,84]

Key elements of the National Academies'Summer Institutes that could be shared as best practices include:

- demonstrate evidence-based teaching methods and engage participants in them as both teachers and learners;
- provide an understanding of the evidence supporting these methods;
- teach participants to use assessment effectively to increase learning and improve teaching;
- provide participants with an opportunity to develop new teaching materials with critical peer review and feedback;
- teach participants how to change their teaching and extend change beyond their own classrooms to foster institutional transformation on their campuses and discipline-wide transformation through their professional societies.

For change in STEM education to become pervasive and propagate independently, a substantial segment of the community needs to be trained so that the language and practice of evidence-based teaching is familiar and embedded in the habits of mind of STEM faculty. Successful programs should be expanded to reach 10-20% of the nation's 230,000 STEM faculty over the next five years. Based on existing teaching training programs, it is reasonable to expect trained faculty to influence the teaching of about 10 colleagues, making it possible to reach a substantial proportion of the STEM faculty through programs targeted at a subset of faculty. Moreover a group consisting of 10-20% of a STEM department's faculty is large enough to become self-sustaining. Such a group is also large enough to handle much or most of the introductory teaching. Therefore, the goal of reaching 10-20% of the STEM faculty directly could result in training most of those who teach in the first two years of college.

A total of $10-15 million per year over 5 years will be required for the training of 23,000 to 46,000 STEM faculty. Funds for this training should be derived from a combination of Federal programs, academic institutions, disciplinary societies, and foundations. For example, funds from NSF's Advanced Technical Education could be used to support training for community college faculty. One possibility is that institutions and private donors could be exhorted to provide funds for this effort through capital campaigns with the theme "building the faculty of the future."

To train future faculty, Federal agencies should require all graduate students and postdoctoral fellows supported by institutional training grants to receive instruction in modern teaching practices. The competencies for postdoctoral researchers developed by the National Postdoctoral Association should receive particular attention.[85] Training could borrow models of success from, for example CIRTL,[86] an NSF-funded program that has created a national network of research campuses that train graduate students and postdoctoral fellows in evidence-based teaching across STEM disciplines. The National Academies Summer Institute model also could be integrated with the CIRTL model, whereby either trainees could attend regional training workshops or directors of campus training programs could be trained centrally and then return to campus to deliver independent teaching workshops. Between 2 and 5% of training grant funds should be set aside, or a supplement of this amount should be added to grants, to provide this instruction.

Using the lever of training grant funding to induce adoption of this training has two important outcomes beyond the students directly affected by the requirement. First, the training is likely to spread beyond the graduate students and postdoctoral fellows who are supported by the training grant. Many graduate students and postdoctoral fellows are eager for this training and will take advantage of it when it is available. Precedent for this is found in the requirement for ethics training instituted by NIH, which rapidly included most graduate students, independent of funding source, at many universities. The second key outcome is that the graduate students at research universities, many of whom are recipients of training grant support, are the future faculty at all types of institutions of higher education. They will therefore become the ambassadors for evidence-based teaching to a wide expanse of colleges and universities.

1-2 Create the "STEM Institutional Transformation Awards" competitive grants program.

NSF should institute a competitive grants program designed to provide incentives for and facilitate transformational, sustainable innovations in the teaching and learning of STEM subjects at two-and four-year colleges. This program could be based on the NSF's ADVANCE Program for increasing the participation of women in STEM[87] or on NSF's Alliances for Graduate Education and the Professoriate, which was designed to increase the participation of minorities. Grants from each of these programs have been successful in effecting transformative change, as established by extensive national studies.[88,89,90] These programs provide the best existing models for institutional level change, which has not historically been a target for Federal funding. The key to these projects is that they focus on *institutional* change and the barriers to it. The interventions developed by each campus should be locally tested but transferrable. The model of the ADVANCE program indicates that a set of model campuses (approximately 100 for ADVANCE) can influence practices at many other campuses by setting an example of successful practices and providing materials that aid other campuses in implementation of similar practices. All ADVANCE projects constructed websites that provided information about program design, data on program impact, and transferrable materials that could be adapted by other campuses. Similarly, a plan to affect other institutions should be part of every STEM Institutional Transformation Award.

The key elements of the award program should include:

- A STEM department's plan to improve education of students in the first two years according to features shown to be important to success of STEM students (see, for example, Table 3)
- Efforts to effect change at the department and institution level
- Sound evaluation to determine whether the interventions have influenced faculty practice and student persistence and performance
- Plan for sustaining programs beyond the duration of the grant
- Evidence of campus commitment to the project through matching institutional funds or other means
- Dissemination of materials to other campuses through websites and publications

Grants also should support putting into practice the large body of existing research on teaching and learning in STEM disciplines by creating incentives for individual departments or entire institutions to adopt or adapt evidence-based methods to improve STEM teaching and learning. The projects should focus on both faculty and department-level reward systems to induce sustainable change. The grants program should allocate $20 million per year over 5 years to fund a total of approximately 100 projects at around $1 million (scaled to the appropriate funding level based on institution type and size). This level of funding would provide sufficient funding for a campus to hire, for example, a teaching expert to assist in development of teaching materials, an evaluator to study faculty behavioral change and effects on student persistence, and a part-time web expert to provide an interface with the central website designed to disseminate materials developed as part of the project. The funding could also be used, in part, to pay for faculty training such as that recommended above. The five-year duration is intended to provide sufficient time to observe change in practices by early adopters, extension of that change to other departments at their institutions, and sharing of progress across institutions through websites and publications. Based on the wildly different capacities and needs at different institution types, grants should be separately considered from community and technical colleges, 4-year colleges, and research institutions. Grants should be awarded based on how proposals address the specific needs of the STEM department or institution and proposed actions that will have the greatest impact on improving student learning and achievement. Funding could come from enactment of NSF's proposed Widening Implementation and Demonstration of Evidence-Based Reforms (WIDER) program at the President's Fiscal Year 2012 requested level of $20 million annually. Project evaluation should focus on changed faculty habits and implementation of evidence-based teaching. In addition, the impacts of teaching practices on retention and on the performance of students in STEM majors should be measured. Granting agencies should not, however, focus on reporting of the effects of interventions on student learning.

Assessment of learning is a standard part of teaching plans, and student persistence and performance (courses taken, grades, time to graduation) should be evaluated, but these measurements should be distinguished from experiments to compare student learning with lectures alone versus with evidence-based methods.

The rationale for this is that evidence-based methods are predicated on research using randomized controlled trials comparing various teaching methods and we do not expect every faculty member teaching an introductory STEM course to perform sophisticated learning science experiments.

It is far more important to document whether and how faculty are implementing the methods. Other sources of funding (such as NSF's Research and Evaluation on Education in Science and Engineering program) could be used to support sound experiments to continue to advance evidence-based learning.

Findings about changed faculty habits and student persistence and performance should be publishable, and materials that are developed should be shared with the academic community through web sites and other means. The grants should transform the campuses receiving them; in turn, these campuses should provide others with models and specific mechanisms for change. The sustainability of change should be planned and evaluated as part of the grant process.

The unique power of educational technology should be harnessed to this end twofold: (1) to embed assessment into instructional activity and use the data gathered to create a virtuous cycle of innovation, sharing, evaluation, and improvement and (2) to disseminate information that can advance the transformation of other instructors, departments, and institutions.

To the latter point, the Department of Education, through the First in the World Initiative or ARPA-Ed (described below under recommendation 2), should issue a request for proposals to produce an interactive, online presence to collect and share data on institutional change and improvement of postsecondary STEM education. Grantees funded by the STEM Institutional Transformation Grants should be required to post their curricula and methods to an online resource.

1-3 Request the National Academies develop metrics to evaluate STEM education.

To evaluate progress toward the goals presented in this report, campuses, funders, students, and accreditation agencies need a meaningful set of criteria by which to measure excellence in STEM teaching among instructors, departments, and institutions. Sufficient research now supports the elements presented here to provide a valid basis for evaluation and benchmarking. The National Science Foundation and Department of Education should provide funding to the National Academies to develop criteria for STEM evaluation based on the partial list provided in Table 3.

Key among these criteria are the capacity to collect, analyze, and use data about teaching and learning; inclusion of effective programs to enhance participation by underrepresented students; incorporation of active student engagement in learning; provision of research experiences in the first two years of college; retention of students in STEM majors; clear demonstration that learning goals guide development of courses and curricula; training in teaching practices for current and future faculty; and evaluation of programs and instructors based on meeting learning goals.

Table 3. Elements of Successful STEM Education Programs

Program Focus	Evidence and Resources
Intellectually engage students	
Teach science with evidencebased methods that engage students in creating and integrating knowledge	Springer, Stanne et al. (1999); AAAS (2011)
Focus on learning goals that involve both the process and content of STEMrelated activities	AAAS (2011)
Involve students in research early, preferably as freshmen	Bartlett (2003); Carter, Mandell et al. (2009) Hathaway, Nagda et al. (2002); Hunter, Laursen et al. (2007); Kight, Gaynor et al. (2006); Kinkel and Henke (2006); Lopatto, Alvarez et al. (2008); Russell, Hancock et al. (2007)
Build alliances between community colleges and research universities to enhance the availability of research experiences to students at community colleges	Shaffer, Alvarez et al. (2010); Wei and Woodin (2011)
Facilitate study group formation and other structures that enable group learning	Burstyn, Sellers et al. (2004); Springer, Stanne et al. (1999)
Personally engage students	
Show relevance of STEM subjects to human and planetary problems	Donofrio, Russell et al. (2007); Buckley, Kershner, et al. (2004)
Provide role models of diverse backgrounds and life choices to inspire diverse students	Lockwood (2006); Stout. Nilanjana et al. (2011); Walton and Cohen (2011)
Provide opportunities for students to become part of STEM communities in classes, research laboratories, and STEMrelated extracurricular activities	Kight, Gaynor et al. (2006); Peckham, Stephenson, et al. (2007)
Show students the diversity of careers in science	Campbell, Fuller et al. (2005)
Provide mentoring and tutoring to help students excel in STEM subjects	Muller (1997); Summers and Hrabowski (2006); Gilmer (2007)
Engage students' families in STEMrelated academic experiences	Rodriguez, GuidoDiBrito et al. (2000); Ong, Phinney et al. (2006); Sy (2008)
Provide students with sufficient resources, including employment in laboratories and scholarships, to enable them to engage fully in academic life and the science and technology community	Barlow and Villarejo (2004)
Provide students with critical feedback and encouragement to give them realistic assessment of their performance in STEM subjects	Ovando (1994)
Build classroom communities in which students feel that they are being groomed for STEM fields rather than weeded out	Gainen (1995)
Build connections between higher education and industry to provide students with internships and exposure to potential career options	Gilmer (2007); Turner, Petzold, et al. (2011)
Provide undergraduate STEM pathways with access to role models by linking graduate training programs with undergraduate research programs	May and Chubin (2003)
Accommodate the needs of nontraditional students	Barlow and Villarejo (2004)
Educate faculty	
Provide faculty with training in teaching through campus programs, summer institutes, and programs organized by professional societies	Pfund, Miller et al. (2009); Yoon, Duncan et al. (2007)
Provide graduate students and postdocs with training in teaching through training grants and professional societies	University of Texas at Austin (2008); BouwmaGearhart (2007); Connolly (2008); Miller, Pfund et al. (2008)
Provide faculty with databases of learning tools and technology	University of Texas at Austin (2011)
Assess outcomes	
Assess understanding through diverse means, and articulate assessment with learning goals	Haudek, Kaplan et al. (2011)

Program Focus	Evidence and Resources
Assess student retention in major	Wild and Ebbers (2002)
Measure achievement gap between various segments of student body and assess impact of interventions on gap	Haak, HilleRisLambers et al. (2011)
Evaluate teaching in terms of learning goals and how they are assessed and met	Felder, Rugarcia et al. (2000)
Improve learning assessment through technology development	Beatty (2004); Caldwell (2007)

Notes: See Appendix I for full references.

These metrics could be adopted by independent organizations, including accreditation agencies, the Association of American Universities (AAU), the Association of Public and Land Grant Universities (APLU), the American Association of Community Colleges (AACC), or U.S. News and World Report, as a way of meaningfully evaluating the quality and success of STEM programs. This effort could be coordinated with ongoing work by AAU[9'] to develop a "STEM Certification" that would be granted to departments that provide outstanding STEM education based on the criteria developed by the National Academies. The inclusion of a STEM education criterion in evaluation of academic departments and institutions will enable prospective faculty and students to make informed judgments and faculty and administrators to benchmark their own progress toward building outstanding STEM undergraduate programs.

When the National Academies develop the undergraduate STEM teaching and learning metrics, they might also consider options for collecting these data, such as the possibility of requiring institutions or STEM departments receiving Federal research funding to report on them. In this case, the responsibility for reporting would be that of the institution or STEM department, not the individual investigator.

Recommendation 2

Advocate and Provide Support for Replacing Standard Laboratory Courses with Discovery-Based Research Courses

Rationale for Recommendation 2

If we taught young people baseball history, statistics, and rules for years before we let them watch or play a game of baseball, how many would become fans or players? Probably few.[92] But in STEM fields, most students must wait until they are quite far along in their studies before then can experience the excitement of scientific research. Solving real-world problems is far more inspiring and instructive than most of the STEM instruction that occurs in the first two years of college. Research experiences in the first two years of college enable students to choose majors based on the best and most creative aspects of STEM fields rather than on courses that do not reflect the nature of inquiry.

Every college student should be given the opportunity to generate scientific knowledge through research. Research experiences in the first two years increase retention of students in STEM majors and improve students'attitudes toward STEM fields.[93,94,95] The effects of research experiences are quite positive for all students but have especially high impact for women and members of other groups currently underrepresented in STEM disciplines (Table 4).

Table 4. Impact of student research experience on students in STEM

Effect	Examples of studies demonstrating effect
Higher grades	Barlow and Villarejo (2004); Junge, Quiñones et al. (2010); Kinkel and Henke (2006)
Identification as a scientist or engineer	Hunter (2007)
Persistence in a STEM major	Barlow and Villarejo (2004); Carter, Mandell et al. (2009); Gilmer (2007); Kinkel and Henke (2006); Summers and Hrabowski (2006)
Shorter time to degree	Kinkel and Henke (2006)
Interest in post graduate education	Foertsch, Alexanmder et al. (1997); Hathaway, Nagda et al. (2002); Kight, Gaynor etal. (2006); Kinkel and Henke (2006); Lopatto (2004); Russell, Hancock at al. (2007)

Notes: See Appendix I for full references.

Not all college students can do research in faculty laboratories. Therefore, we propose the extensive use of research courses, which have been successfully implemented and rigorously studied at both large and small institutions and have been proven to increase student knowledge of, enthusiasm for, and retention in STEM (e.g., see Appendix G).[96,97]

Two strategies to engage students in research in their first two years are necessary: (i) widespread integration of research courses into the introductory STEM curricula, and (ii) increased opportunities for students to participate in faculty research programs in the first two years.

Actions to Achieve Recommendation 2

2-1 Expand the use of scientific research and engineering design courses in the first two years of postsecondary education through an NSF program.

All available data show that traditional cookbook, introductory laboratory courses which often involve repeating classical experiment to reproduce known results, rather than engaging students in experiments with the possibility of true discovery produce less learning, inspiration, and retention in STEM disciplines than do research courses. The data suggest that approaches to the development and scale-up of research courses should be made available through current and future faculty training programs and centralized websites and could be an important component of STEM Institutional Transformation Awards and STEM Certification (Recommendation 1-2). Research courses can act as training for subsequent participation in research in faculty or industry laboratories, improving the skills that students bring to those positions.

The National Science Foundation should provide initial funding to replicate and scale-up model research or design courses, possibly through the existing Transforming Undergraduate Education in STEM (TUES) program or the Science, Technology, Engineering, and Mathematics Talent Expansion Program (STEP). On the order of thirty percent of the existing programs across STEM disciplines could be focused on funding for implementing research courses at postsecondary academic institutions at an annual cost of approximately $12.5 million dollars, annually (based on Fiscal Year 2010 funding levels). Based on the range of funding for Type 3 TUES grants and Type 1 STEP grants, about 10 proposals per year at an average level of $1.2 million could be awarded, in order to impact 100 campuses over the next 10 years.

Colleges and universities should seek to match NSF funding with private and philanthropic sources. Because research courses will replace expensive introductory laboratory courses, they should not require ongoing external support once the transition is accomplished.

2-2 Expand opportunities for student research and design in faculty research laboratories by reducing restrictions on Federal research funds and redefining a Department of Education program.

Independent research on faculty projects is a direct way for students to experience real discovery and innovation and to be inspired by STEM subjects.

All relevant Federal agencies should rigorously examine their programs which support undergraduate research and where there exists prohibitions, either in policy or practice, which would interfere with the recommendations of this report to support early engagement of students in research, these should be changed. Novel opportunities for research areas—for example, in theoretical areas without laboratories — should also be supported as arenas for involvement by students in the first two years.

Federal agencies should give special consideration to proposals for Federal training grants that establish collaborations between research universities and community colleges or other institutions that lack faculty research programs, including minority serving institutions that fall into this category.

In these programs, graduate students supported on training grants would learn to be effective research mentors and have the opportunity to work with undergraduates from other institutions.

Potential sources of funding include redefinition of the Department of Education's $1 billion Carl D. Perkins Career and Technical Education program when it comes up for reauthorization; specific focusing of Federal (NASA, Energy, Education, and USDA) investments in historically black colleges and universities, Hispanic-serving institutions, and tribal colleges for building institutional capacity; and NSF's Broadening Participation at the Core on supporting early research and building cross-institutional collaborations in undergraduate research.

Recommendation 3

Launch a National Experiment in Postsecondary Mathematics Education to Address the Mathematics-Preparation Gap

Rationale for Recommendation 3

Students need mathematical and, increasingly, computational competency at a college level to succeed in STEM majors and jobs. This makes mathematics distinct from other disciplines, as it is a gateway to other STEM fields.

Mathematics instruction credit hours, particularly in the first two years, are dominated by "service courses"—mathematics courses that are taken because they are required by another major that uses mathematics in the discipline. This content is fundamentally different from how a pure mathematician thinks about mathematics or knows how to use it, which is problematic for teaching students the skills they need. Discipline-based education on effective undergraduate mathematics teaching also appears less developed when compared with other STEM fields.

Today, many students entering college do not meet the necessary mathematics standards. Among students who take the ACT entrance examination for college, just 43% achieve the

ACT College Readiness Benchmark in math.[98] Because of inadequate preparation, many students need to take developmental classes in mathematics when they get to college. In addition, employers in the private sector, government, and military frequently need employees with a level of mathematics preparation that is hard to find, placing the burden on employers to provide or obtain remedial education. This deficiency in mathematics skills imposes a burden on students, higher education, the military, and employers through the developmental education and worker training needed to produce a STEM competent workforce. Higher education alone spends at least $2 billion dollars per year on developmental education.[99] This high cost for remediation is coupled to reported low effectiveness. Of those students who took remedial courses, less than 30% complete a bachelor's degree within eight years as compared to nearly 60% of students who required no remediation.[100] Additionally, a study of community college students reported that 60-70% of students who were placed in remedial mathematics never completed the required mathematics sequence and, therefore, never graduated.[101] Reducing or eliminating the need for remedial mathematics classes or improving their cost and effectiveness is one of the most urgent challenges—and promising opportunities—in preparing the STEM workforce of the 21st century.

Undergraduate mathematics education in the U.S. is often below what is considered the appropriate university level in many countries. In 2005, 57% of the students enrolled in 4-year colleges and universities were enrolled in pre-college algebra, trigonometry, or other pre-calculus courses; the proportion is higher for 2-year institutions. Most U.S postsecondary students terminate their college mathematics education at a pre-calculus course that is typically a review of high school algebra, trigonometry, and sometimes functions. Many students in these courses have seen 90% of the material in high school but are advised to take this course to make the transition to college easier.[102] Such courses are frequently uninspiring, relying on memorization and rote learning while avoiding richer mathematical ideas.

As this is the last mathematics course for many college students, they often are left with the impression that the field is dull and unimaginative, and they can extend this judgment to all STEM disciplines. This is particularly harmful for students who later become K-12 STEM teachers. A focus on improving this particular type of course offers potentially enormous benefits.

Closing the mathematics-preparation gap would enable many more students to pursue STEM degrees in college. About 15% of 12th graders are interested in STEM fields but not proficient in mathematics, with women slightly more common in this category. Many of these students are not far from mathematics proficiency (see Appendix E). If the preparation of these students in mathematics could be enhanced, either before college enrollment or through improved remediation, many more students could be prepared to pursue STEM fields in college.

This problem is a complex one that has resisted the efforts of many dedicated people over a considerable period of time. Closing the mathematics-preparation gap requires coordinated action on many fronts starting in the earliest grades.[103] PCAST's report on K-12 STEM Education ("*Prepare and Inspire*"[104]) contains several recommendations that involve colleges and universities in this effort. In particular, it calls for the Federal Government to establish the objective of recruiting, preparing, and providing induction support for at least 100,000 new middle and high school STEM teachers who have majors in STEM fields and strong content-specific pedagogical preparation. This Administration has embraced this goal, and the production of 1 million additional STEM graduates over the next decade could contribute substantially to meeting it.

Secondly, the Federal Government has a critical role in supporting the development of a knowledge base to close the mathematics-preparation gap. For example, research into the best ways to teach mathematics to college students so they can pursue STEM subjects in the first two years of college is badly needed. Some developmental mathematics courses have demonstrated effectiveness in increasing mathematics proficiency among those not ready for college-level mathematics and even in encouraging students intending to major in STEM subjects to persist to graduation and a STEM degree.[105] Mathematics education research should explore the attributes of these successful classes and ways to disseminate best practices.

Mathematics education research also could lead to innovative and effective ways to teach the subject— for example, through the use of games, simulations, and other technologies. Emerging computer-based technologies—intelligent tutors—based on the latest learning science hold promise for accelerating mathematics learning and achieving mathematics proficiency at less cost than current approaches (see Appendix E, Box E-1). Preliminary research suggests that intelligent tutors can increase mathematics test scores on the order of one to two standard deviations. Further development and broad diffusion of these tools can provide effective, low-cost strategies for accelerating mathematics learning among STEM-interested students.

In the *Prepare and Inspire* report, PCAST called for the creation of a mission-driven, advanced research projects agency for education (ARPA-Ed) that would propel and support (1) the development of innovative technologies and technology platforms for learning, teaching, and assessment across all subjects and ages, and (2) the development of effective, integrated, whole-course materials for STEM education. Many of these advances would benefit not only K-12 education but also the developmental courses that many students need to pursue STEM fields during the first two years of college.

Actions to Achieve Recommendation 3

3-1 Support a national experiment in mathematics undergraduate education at NSF, the Department of Labor, and the Department of Education.

The National Science Foundation and the Departments of Labor and Education should support a multi-campus five-year initiative aimed at developing new approaches to remove or reduce the mathematics bottleneck that is currently keeping many students from pursuing STEM majors.

This national experiment should fund a variety of approaches, including

(1) summer and other bridge programs for high school students entering college;

(2) remedial courses for students in college, including approaches that rely on computer technology;

(3) college mathematics teaching and curricula developed and taught by faculty from mathematics-intensive disciplines other than mathematics, including physics, engineering, and computer science; and

(4) a new pathway for producing K-12 mathematics teachers from undergraduate and graduate programs in mathematics-intensive fields other than mathematics.

Diverse institutions should be included in the experiment to assess the impact of the intervention on various types of students and schools.

Outcome evaluations should be designed as a collective effort by the participating campuses and funding agencies.

Approximately 200 experiments at an average level of $500,000 should be funded at institutions across the county, at an annual cost of $20 million per year for five years.

As mathematics preparation issues vary across the postsecondary spectrum, a variety of sources will be needed to fund experiments at diverse institution types.

Funds for these experiments could be derived from a combination of the Department of Education's proposed First in the World Initiative, possibly the Department of Labor's Trade Adjustment Assistance Community College and Career Training initiative or Career Pathways Innovation Fund, and a strategic focus on mathematics of NSF's Transforming Undergraduate Education in STEM (TUES) program or Science, Technology, Engineering, and Mathematics Talent Expansion Program (STEP) for the next five years.

Recommendation 4

Encourage Partnerships among Stakeholders to Diversify Pathways to STEM Careers

Rationale for Recommendation 4

Besides increasing student persistence in STEM education, more students need to be attracted to STEM disciplines to produce 1 million additional college graduates over the next decade. To take advantage of the breadth of talent in STEM fields, students who need non-traditional pathways to STEM degrees require special attention. Adult and working students and those from backgrounds atypical of traditional STEM students, including returning veterans, may need alternative pathways to be successful in STEM disciplines.

New STEM pathways need to offer nationally portable, industry recognized credentials that are integrated into for-credit academic degree programs. These programs provide bridges from high school to community colleges, from community colleges to 4-year institutions, and from all institution types to STEM jobs.

The sizeable group of high school dropouts who return to study for General Education Development (GED) tests offers a largely untapped source of students who could be interested in careers involving STEM fields. Some community colleges have begun offering programs that combine preparation for the GED tests with college courses that could serve as a gateway to further STEM courses and STEM-related careers. Adult students and those returning to college after time away, especially U.S. military veterans, also often have high levels of motivation and a focus on careers that could be channeled in the direction of STEM-related jobs.

Educators concerned with increasing the number of students in STEM disciplines have given much attention to "off-ramps", the drop-out and attrition patterns in high school, community colleges, and four-year institutions. Equal attention should be given to on-ramps, multiple routes to enter or re-enter STEM education. Rather than a single pipeline that is prone to leakage, or a ladder where any missed step makes the next step too hard to reach, educators and policymakers should think of a network of pathways along which students can take different routes to STEM readiness and competency. If students have exited this network of pathways, they need accessible and cost-effective ways to get back on.

Many types of partnerships could aid in designing pathways to STEM training that would capture a broader portion of society. These partnerships can smooth the way from high school to college, link students at community and technical colleges with high-skill STEM jobs, enable students at two-year colleges to transfer and earn four-year degrees, provide research experiences to students at institutions without research programs, and offer students insight into the careers and opportunities for STEM practitioners in industry. These partnerships will enable the academic advancement of all types of students, but they will be particularly advantageous for students traditionally underrepresented in STEM fields.

Partnerships between High School and College

To encourage the underrepresented majority to pursue STEM degrees, better integration between high school and college is needed. High schools and colleges could collaborate on development of bridge programs that would prepare students for college during the summer between high school and college (e.g., Appendix H, Box H-1 and H-2). Typically, high school juniors and seniors in these programs live on campus and receive classroom instruction, research experience, career counseling, SAT and ACT prep classes, and mentoring from students and faculty.

Most of these programs, such as Carnegie Mellon University's Summer Academy for Mathematics and Science[106] and the California State Summer School for Mathematics and Science,[107] are open to high school students statewide or nationwide. Some are aimed at the underrepresented majority to provide incoming students with the intellectual, personal, and social supports they will need to excel.

The majority of partnerships between high schools and colleges and universities that aim to increase the number of students entering STEM pathways do so indirectly by providing better teacher training and support, which in turn can lead to more students interested in STEM disciplines and better prepared to enter college. Such programs can train high school teachers to use new tools for active learning that engage students in hands-on STEM activities. These programs also can provide on-site coaching and leadership development for principals and other administrators.

Partnerships between Two- and Four-Year Institutions

Two-year colleges are both a major source of STEM degrees and a conduit into STEM fields for many students, including many members of the underrepresented majority.[108] In many cases, 2-year colleges are geographically closer to students who would have difficulty attending a four-year school. They also are typically less costly than 4-year institutions.[109]

The transition from a 2-to 4-year college can be difficult, especially for members of groups underrepresented in STEM fields. For example, a recent study of women of color in STEM disciplines found that transfer rates are low for women of color, and retention rates of transfer students in STEM programs at 4-year institutions are even lower.[110] Programs that align credentials between 2- and 4-year schools and provide support through the transfer process can increase the success rate for students in STEM fields. Collaborative partnerships between 2-year colleges and 4-year institutions would provide greater access to and opportunities for advanced STEM education (see Appendix H, Box H-4) and advanced training for those students who do not pursue a 4-year degree. Large state systems, such as the

University of California and California State University systems, have long-standing programs like MESA (Mathematics, Engineering, Science Achievement) that create partnerships between 4-year universities and neighboring 2-year colleges to align curricula and work with students to ensure that they are well-prepared for the transition to bachelor's degree programs in STEM disciplines. Courses at the community college vetted by university faculty not only provide the necessary intellectual rigor but also allow students to develop relationships with faculty prior to transferring. In addition, students enrolled in these programs can be granted access to libraries and can be provided with opportunities to participate in cultural and athletic events at the university, helping them more easily integrate into campus life upon successful transfer.

Partnerships Involving Minority-Serving Institutions

Minority-serving institutions (MSIs) can serve as key intermediaries to improve the numbers, preparation, and diversity of students interested in STEM fields.[111,112] For example, through a combination of bridge programs, building a community of support for STEM students, increasing student research opportunities, and reevaluating teaching and research practices, the University of Texas-El Paso has boosted graduation rates in STEM disciplines by nearly 50%, transforming it into the largest producer of Mexican-American STEM graduates in the Nation.[113] Several White House initiatives have directed funds to MSIs to increase the number of minority students who not only start but finish STEM degree programs.[114] Collaborative efforts between MSIs and other colleges and universities could greatly improve educational experiences in STEM disciplines. Programs that enhance STEM curricula at MSIs and that focus on improving the readiness of first-year students, often through summer research experiences and laboratory experiences at partnering research universities, have shown marked success at both increasing enrollment and retention of students in STEM disciplines (see Appendix H, Box H-5).

Partnerships between the Private Sector and Undergraduate STEM Education

Many U.S. businesses are active supporters of STEM efforts in high schools, colleges, and universities. Recently, however, the U.S. business community has recognized that its traditional role of partnering with existing institutions to promote best practices, provide resources, and involve corporate supporters offers some aid but is not likely to produce the radical change needed to meet future STEM workforce needs.[115] Providing mentoring for promising STEM students through cooperative education, learn and earn, and internship programs is an important and proven avenue by which businesses can both recruit future workers and help students complete their studies. Not only do cooperative education experiences provide the kind of hands-on training shown to increase student retention in STEM programs, but they also produce students who more quickly integrate into the workplace and express higher rates of job satisfaction (e.g., Appendix H, Boxes H-6 and H-7). There are several ongoing efforts in this area at the Federal level focused on distinct aspects of the STEM workforce, and these should be incorporated into a broader strategy for partnerships between industry and institutions of higher education to improve engagement and training of undergraduate students in STEM disciplines. For example, the President's Council on Jobs and Competitiveness[116] has set a goal to double internship offerings from partnering businesses to increase the supply of qualified and trained American engineers.[117] In addition, the Advanced Manufacturing Partnership,[118] launched by President Obama in

June 2011 in response to recommendations from a PCAST report on the topic,[119] has a workstream on education and workforce demand that is exploring opportunities for partnerships between employers and educational institutions, with a particular focus on community colleges but also including high schools and universities. An untraditional avenue by which businesses can partner with universities to bolster pathways to STEM careers is to help transitioning employees become STEM educators. IBM launched such a program in 2005[120] and has helped over 100 employees start new careers in teaching. The program works with university partners to provide employees who have engineering, health, computing, and science backgrounds with the resources needed to custom-tailor their own teacher preparation through on-line and traditional classes while retaining their jobs at IBM until their training is complete.

Actions to Achieve Recommendation 4

Increasing the number and strength of pathways to STEM fields during the first two years of college requires a coordinated strategy involving not just the Federal Government but K-12 education, higher education, businesses, and the nonprofit sector, including foundations. Still, the Federal Government can lead this effort through strategic planning, reallocation of funding, and strong leadership.

4-1 Sponsor at the Department of Education summer STEM learning programs for high school students.

The Department of Education should roll out the summer learning programs authorized in the 2007 America Competes Act (in an amendment introduced by then-Senator Obama), funded in part by the Federal Government, in partnership with state and local entities. To cover the full costs of such programs, state and local entities should recruit institutions of higher education and private industry as partners. As an expansion of the original proposal, these programs should be focused on mathematics, engineering, and science for high school students. In particular, these programs should provide mathematics instruction to prepare students to be college- and career-ready and provide hands-on STEM experiences. Based on the size of National Science Foundation's former Young Scholars Program for summer institutes, we recommend investment of $10 million to fund approximately 100 projects reaching on the order of 5000 students, annually, including significant cost-sharing with academic institutions and private investors. In addition, as authorized in the law, the Department of Education should establish a "Summer Learning Grants Website" that would provide information for students, parents, and educators on successful programs, curricula, and best practices for summer learning opportunities.

4-2 Expand a Department of Labor Program scope to encourage pathways from two-year to four-year institutions.

The mission of the Department of Labor's Trade Adjustment Assistance Community College and Career Training initiative should be expanded beyond development of important partnerships between community and technical colleges and employers in the private sector to encourage scientific research and engineering design exchanges across 2- and 4-year institutions. Alternatively, these activities could be funded through a strategic focus of the Department of Labor's Career Pathways Innovation Fund on research partnerships. NSF's Advancing Technical Education program could also be focused on cross institutional collaborations.

The bridges described here should provide authentic STEM experiences for community college students on the 4-year campus and allow them to develop relations with faculty and the college or university community to ease the potential transition from a 2- to 4-year institution or to provide advanced experiences and inspiration for students who do not pursue a 4-year degree.

4-3 Establish public-private partnerships to support successful STEM programs.

To enhance students'STEM readiness, the Federal Government should engage private industry and foundations to support successful programs that create bridges between high schools and colleges and between 2- and 4-year institutions and ensure that programs incorporate learning standards and content consistent with industry-recognized skills.

In the model of Change the Equation, for which business leaders stood with President Obama and committed financial investment in strategies that work in K-12 STEM education, the President should call on foundations and private industry to commit to improving recruitment and retention in undergraduate STEM education and to partner with Federal agencies to expand education technologies, provide internships to students in the first two years of college, and invest in programs with proven success (such as cohort programs, bridge programs, and certification programs linking community college and technical education to industry-recognized standards).

Particular attention should be paid to U.S. military veterans who often have exceptional levels of motivation, maturity, and focus along with STEM skills gained during their service.

Defense-related industries should consider partnerships with the Department of Defense and Veterans Affairs to support efforts to train and certify veterans for careers in STEM and STEM-capable fields.

This commitment could involve industry offering internships and learn and earn programs to veterans who enroll in college to enhance their workplace experience and improve their job-readiness upon graduation.

4-4 Improve data provided by the Department of Education and the Bureau of Labor Statistics to STEM students, parents, and the greater community on STEM disciplines and the labor market.

The private sector and the Federal agencies that run laboratories and employ STEM professionals and the STEM-capable workforce have a vested interest in high-quality information about effective STEM education and relevant data about workforce supply, demand, and skill levels.

Current data sources, however, limit their ability to answer important questions about the skills and choices of workers and about trends in the supply of and demand for a STEM and STEM-capable workforce.

One way to help mitigate this data gap would be to resurvey members of cohorts followed in the High School and Beyond and National Education Longitudinal studies.

Also, the National Center for Education Statistics within the Department of Education's Institute of Education Sciences should facilitate the enhancement of state student unit record systems to permit matching to postsecondary school data and labor market outcomes.

The Bureau of Labor Statistics should redefine employment categories around STEM jobs to reflect the breadth of jobs that require STEM skills, including STEM-capable jobs, such as medical and advanced manufacturing professionals and K-12 STEM educators.

Recommendation 5

Create a Presidential Council on STEM Education with Leadership from the Academic and Business Communities to Provide Strategic Leadership for Transformative and Sustainable Change in STEM Undergraduate Education

Rationale for Recommendation 5

The leadership of higher education and STEM-dependent industries needs to be inspired to generate sweeping change in higher education to produce the workforce America needs. The leaders in these sectors need to be challenged by the country's political leaders to think creatively, design and implement programs, to challenge existing reward structures, and to raise money from private donors to benefit STEM education. The White House should add its voice to this cause to help these leaders take charge of STEM education in their institutions and lead the way to new levels of achievement in STEM education.

Actions to Achieve Recommendation 5

The leadership of higher education and STEM-enabled businesses needs to be inspired to generate sweeping changes in higher education to produce the workforce America needs. Toward this end, we recommend that the President, via Executive Order, form a Presidential Council on STEM Education to provide advice and leadership on postsecondary STEM education. The council should include members that represent the breadth of academic institutions, professional societies, businesses, and private foundations involved in the development and use of human capital in STEM fields. Based on the guidance provided in this report, the council should make recommendations that advance the quality of postsecondary STEM education through all mechanisms available to the President. The council could provide a forum for leaders in the public and private sectors to weigh in on the development and deployment of metrics to evaluate STEM departments (Recommendation 1) and to design collaborative coalitions to support initiatives in STEM education (Recommendation 4), including expanding internship programs in industry and connecting industrial research agendas with research courses (Recommendation 2). In addition, it could provide advice and review for the National Experiment in Mathematics Undergraduate Education (Recommendation 3) and could conduct further study of the mathematics education issue, if necessary.

V. Engage to Excel: Summary of Recommendations, Actions, and Estimated Costs

Table 5. Engage to Excel: Summary of Recommendations, Actions, and Estimated Costs

1. Catalyze widespread adoption of empirically validated teaching practices.	
Action	Agency and Estimated Cost
Establish disciplinefocused programs funded by Federal agencies, academic institutions, professional societies, and foundations to train (1) current and (2) future faculty in evidencebased teaching practices.	1.NSF and other agencies should partner with foundations and disciplinary societies to expand existing teacher training programs ($10$15 M per year over five years to train 23,000 to 46,000 STEM faculty).

Table 5. (Continued)

1. Catalyze widespread adoption of empirically validated teaching practices.	
Action	Agency and Estimated Cost
	2.All agencies that provide training grants for graduate students and postdocs, through a combination of training grants and institutionalfunds.
(1) Create a "STEM Institutional Transformation Awards" competitive grants program at NSF. (2) Develop an online presence to share data and best practices.	1.NSF's proposed Widening Implementation and Demonstration of Evidencebased Reforms (WIDER) program. $20 M per year over fiveyears to fund 100 multiyear projects. 2.Education through proposed First in the World Initiative or ARPAEd.
Request that the National Academies develop metrics to evaluate STEM education.	NSF and Education to request this study, with cost to be determined.
2. Advocate and provide support for replacing standard laboratory courses with discovery-based research courses.	
Action	Agency and Estimated Cost
Expand the use of scientific research and engineering design courses in the first two years through an NSF program.	NSF, with initial funding possibly through Transforming Undergraduate Education in Science (TUES) or Science, Technology, Engineering, and Mathematics Talent Expansion Program (STEP) at $12.5 M, annually (i.e. 10 Type 3 TUES or Type 1 STEP proposals per year at an average of $1.2M).
Expand opportunities for student research in faculty laboratories by (1) reducing restrictions on Federal research funds, (2) giving special consideration to training grants that establish collaborations between research universities and other institutions, and (3) redefining a Department of Education program.	1.All Federal agencies should make it possible to use undergraduate research program funds for firstand secondyear students. 2.Federal agencies that fund programs for minority institutions could encourage crossinstitution research partnerships. 3.Include research opportunities as technical education, such as that supported by the Department of Education's Carl D. Perkins CTE program.
Support a national experiment in mathematics undergraduate education focused on: (1) summer programs; (2) remedial courses including use of technology; (3) disciplinebased mathematics instruction, and (4) new pathways for K12 mathematics teachers.	Fund 200 sites at an average of $500,000 over five years, or $20 M per year for five years, with funds from: NSF's TUES or STEP programs, DOL's Trade Adjustment Assistance Community College and Career Training (TAACCCT) Grant Program or Career Pathways Innovation Fund, and Education's proposed First in the World Initiative.
4. Encourage partnerships among stakeholders to diversify pathways to STEM careers.	
Action	Agency and Estimated Cost
Sponsor summer STEM learning programs for high school students.	Education as authorized in the America Competes Act ($10m to fund about 100 projects reaching on the order of 5000 students, annually).
Expand the scope of a DOL program and focus an NSF program to encourage pathways from 24 year institutions.	DOL's TAACCCT Grant Program initiative or Career Pathways InnovationFund or NSF's Advancing Technical Education program to support community collegeuniversity or college research and design partnerships.
Establish publicprivate AgencyInstitutionIndustry partnerships to support successful STEM programs.	All STEM and educationfocused Federal agencies.
Improve data provided to STEM students, parents, and the greater community on STEM education disciplines and the labor market.	Department of Education should devote more resources to tracking students from high school into their careers. Bureau of Labor Statistics should redefine employment categories to include in "STEM" the breadth of jobs that require STEM skills.
5. Create a Presidential Council on STEM Education with leadership from the academic and business communities to provide strategic leadership for transformative and sustainable change in STEM undergraduate education.	

APPENDIX A: EXPERTS PROVIDING INPUT TO PCAST

PCAST and its STEM Undergraduate Education Working Group express gratitude to the following individuals, who contributed input by attending meetings or by responding to requests for information:

Joann Boughman
Chief Executive Officer
American Society of Human Genetics

David Burgess
Professor of Biology
Boston College
Board Member, Treasurer, and
Past-President Society for the Advancement
of Chicanos and Native Americans in Science

Ashwin Ram
Director, Cognitive Computing Lab and Associate
Professor, College of Computing
Georgia Tech
Founder
Enkia Corporation

Cliff Adelman
Senior Associate
Institute for Higher Education Policy

Bruce Alberts
Editor-in-Chief *Science*

Mary Ann Rankin
Dean, College of Natural Sciences
University of Texas at Austin

Joseph Aoun
President
Northeastern University

David Asai
Director of the Precollege and
Undergraduate Science Education Programs
Howard Hughes Medical Institute

Margaret Ashida
Project Director
Empire State STEM Education Initiative
Rensselaer Polytechnic Institute (RPI)

Debbie Bial
President
The Posse Foundation

Board on Science Education (BOSE)
Division of Behavioral and Social Science and Education,
National Research Council The National Academies

Kenneth Boutte, Sr.
Professor of Biology and Associate
Dean of Summer
Programs and External Initiatives
Xavier University of Louisiana

Michelle Cahill
Vice President for National Programs Carnegie
Corporation of New York

Anthony Carnevale
Research Professor and Director,
Georgetown University Center on
Education and the Workforce
Georgetown University

Julie Carruthers
Senior Science and Technology Advisor
Office of the Deputy Director of Science Programs
U.S. Department of Energy

David Conley
Professor of Educational Policy and Leaders,
College of Education
University of Oregon

Sandra Everett
Technical Projects and Training Manager
Lorain County Community College Nord
Advanced Technology Center

John Ewing
President
Math for America

Joan Ferrini-Mundy
Assistant Director for Education and
Human Resources
National Science Foundation

Norman Fortenberry
Executive Director
American Society for Engineering Education

Richard Freeman
Herbert Ascherman Chair of
Economics Harvard University

Jeff Froyd
Director of Academic Development, Dwight Look
College of Engineering
Project Director, NSF Foundation Coalition
Texas A&M University

William Gary
Vice President for Workforce
Development Northern Virginia Community College

Marybeth Gasman
Professor
University of Pennsylvania Graduate
School of Education

Howard Gobstein
Executive Vice President for Research,
Innovation, and STEM Education
Association of Public and Land Grant
Universities Co-Director
Science and Mathematics
Teaching Imperative

William Goggin
Executive Director
Federal Advisory Committee on Student
Financial Assistance
U.S. Department of Education

Gary Green
President
Forsyth Technical Community College

Peter Henderson
Director
Board on Higher Education and Workforce
National Research Council
The National Academies

Gary Hoachlander
President
ConnectEd: The California
Center for College and Career

Freeman Hrabowski
President
University of Maryland,
Baltimore County

Raynard Kington
President
Grinnell College

Kate Kirby
Executive Officer
American Physical Society

Mary Kirchhoff
Director
American Chemical Society
Education Division

Ken Koedinger
Director
Pittsburgh Science of Learning
Center Professor
Human-Computer Interaction
Institute Carnegie Mellon University

Dennis Kox
Manager
Advanced Manufacturing
Raytheon Missile Systems

Michael Lach‡
Special Assistant
U.S. Department of Education

Alan Leshner
Chief Executive Officer
American Association for the Advancement of Science
Executive Publisher
Science

James Lightbourne
Senior Advisor
Directorate for Education and
Human Resources National Science Foundation

Anne MacLoughlin
Senior Researcher
Center for Studies in Higher Education
University of California, Berkeley

Cathy Manduca
Director
Science Education Resource Center
Carleton College
Robert Mathieu
Professor of Astronomy
University of Wisconsin-Madison
Director
Center for the Integration of Research,
Teaching, and Learning (CIRTL)

Krish Mathur
STEM Liason
Office of Post Secondary Education
Program Officer
Fund for the Improvement
of Postsecondary
Education
U.S. Department of Education

Irving McPhail
President and CEO
National Action Council for Minorities in Education

‡ As of November 2011, Michael Lach serves as Director of STEM Education at the Urban Education Institute, University of Chicago.

Harris Miller
CEO and President
Association of Private Sector Colleges
and Universities (APSCU)

Mitzi Montoya
Vice Provost and Dean,
College of Technology & Innovation
Chair
Department of Technological
Entrepreneurship & Innovation Management
Arizona State University

Jeanne Narum
Director
Independent Colleges Office
Founding Director
Project Kaleidoscope
Patrick Natale
Executive Director, Chief Staff Officer and Secretary
American Society of Civil Engineers

Ben Noel
President and Chief Executive
Officer 360Ed

Willard Nott
Vice President for Public Awareness
American Society of Mechanical Engineers

Jane Oates
Assistant Secretary for Employment and
Training U.S. Department of Labor

Eduardo Ochoa
Assistant Secretary for Postsecondary
Education U.S. Department of Education

Michael Pearson
Associate Executive Director
Mathematical Association of America

Jan Plass
Co-Director
NYU Games for Learning Institute (G4LI)

Founding Director
CREATE
Professor of Educational '
Communication and Technology
New York University

Clifton Poodry
Director
Division of Minority Opportunities in
Research National Institutes of Health

Samuel Rankin, III
Associate Executive Director American
Mathematical Society

Mark Regets
Senior Program Officer
Board on Higher Education Workforce
(BHEW) National Research Council
The National Academies Senior Analyst Science
and Engineering Indicators Program Division
of Science Resources Statistics National Science Foundation

Linda Roberts
Former Director
Office of Education Technology
U.S. Department of Education

Steven Robinson
Special Assistant Office of Elementary
and Secondary Education White House
Domestic Policy Council

Deborah Santiago
Co-founder and Vice-President for
Policy and Research Excelencia in Education
Victor Santiago
Deputy Division Director
Division of Human Resource Development
National Science Foundation

Martin Scaglione
President and COO
Workforce Development Division
ACT, Inc.

Mel Schiavelli
President
Harrisburg University of
Science & Technology

Betty Shanahan
Executive Director and Chief Executive
Officer Society of Women Engineers

Tobin Smith
Vice President for Policy
Association of American Universities

Blair Smith
Dean
College of Information Systems and
Technology University of Phoenix

Michael Staton
Chief Executive Officer and Co-Founder Inigral, Inc.

Martin Storksdieck
Director
Board on Science Education
National Research Council
The National Academies

Vincent Tinto
Distinguished University Professor
Former Chair
Higher Education Program Syracuse University

Mark Tomlinson
Executive Director and General Manager
Society of Manufacturing Engineers

Josh Trapani
Senior Policy Analyst
Association of American Universities

Uri Treisman
Professor of Mathematics and Public Affairs
University of Texas at Austin
Founder and Director
Charles A. Dana Center

Jodi Wesemann
Treasurer and Member of the Board
Association for Women in Science

John Wilson
Executive Director
White House Initiative on Historically
Black Colleges and Universities

Michael Wolf
Supervisory Economist
Employment Projections Program
Bureau of Labor Statistics

Bill Wood
Co-Director
National Academies Summer Institute on
Undergraduate Education in Biology
Professor, Molecular, Cellular,
and Developmental Biology
University of Colorado at Boulder

Frederico Zaragoza
Vice Chancellor of Economic
and Workforce Development
Alamo Colleges

Rodney Ulane
NIH Training Officer
Director, Division of
Scientific Programs
Office of Extramural Research
National Institutes of Health

APPENDIX B : ACKNOWLEDGMENTS

PCAST wishes to express gratitude to the following individuals who contributed in various ways to the preparation of this report:

Asha Balakrishnan
Research Staff Member
Science and Technology
Policy Institute

Greg Gershuny[Π]
Confidential Assistant White House
Office of Science and Technology Policy

Michelle Costanzo
Student Volunteer PCAST

Michael Feder
Policy Analyst White House Office of
Science and Technology Policy

Dara Fisher
Student Volunteer
White House Office of Science
and Technology Policy
Kumar Garg
Policy Analyst White House Office of
Science and Technology Policy

Gera Jochum
Policy Analyst White House Office of
Science and Technology Policy

Thomas Kalil
Deputy Director for Policy
White House Office of Science
and Technology Policy

Brandon Ledford
Student Volunteer PCAST

Elizabeth Lee
Research Assistant
Science and Technology Policy Institute

Justin Scott
Research Staff Member
Science and Technology Policy Institute

Sam Thomas Research Assistant
Science and Technology Policy Institute

Π Greg Gershuny left OSTP to join the White House Office of Presidential Personnel as the Energy and Environment Director in October 2011.

Maya Uppaluru
Student Volunteer
White House Office
of Science and Technology Policy

Carl Wieman
Associate Director for Science
White House Office of Science
and Technology Policy

APPENDIX C: STEM HIGHER EDUCATION ENROLLMENT, PERSISTENCE, AND COMPLETION DATA

Appendix C describes data regarding the STEM enrollment, persistence and completion for students studying at postsecondary institutions.

This appendix is the result of work done by Institute for Defense Analyses' (IDA) Science and Technology Policy Institute (STPI).

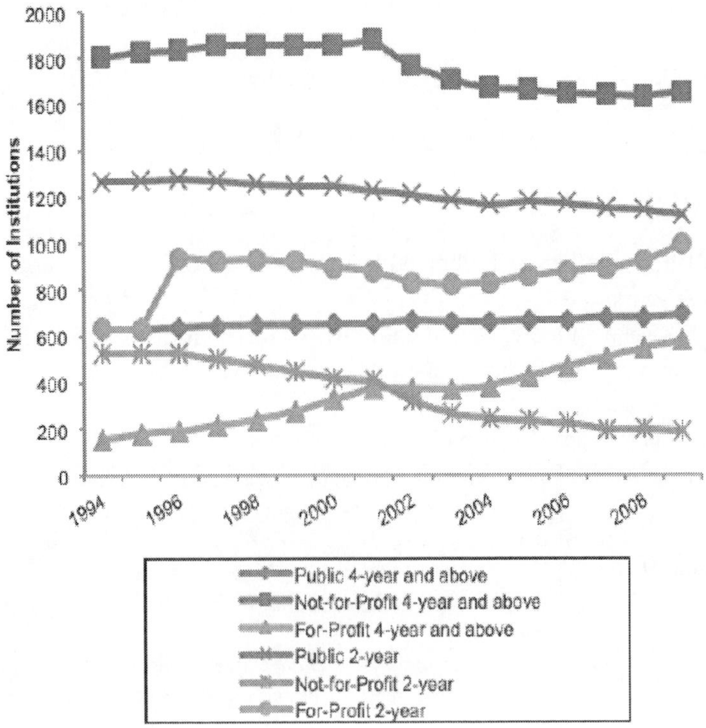

Derived from: IPEDS, National Center for Education Statistics.

Figure C-1. Number of Institutions by Sector, 1994 to 2009.

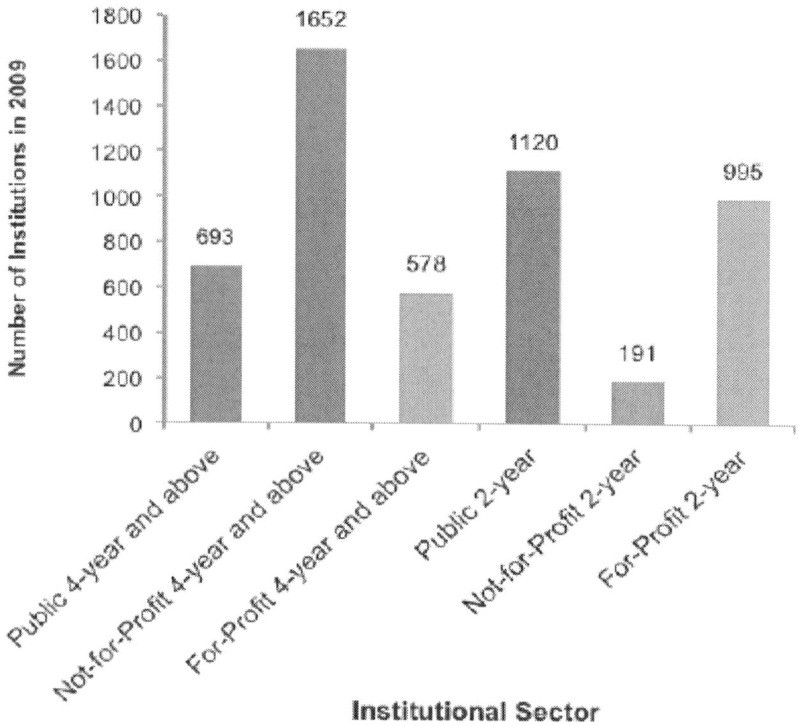

Derived from: IPEDS, National Center for Education Statistics.

Figure C-2. Number of Institutions by Sector Breakdown, 2009.

Key Points (Figure C-1 and Figure C-2):

- Overall, the number of higher education institutions has remained steady over the past five years.
- There has been an increase in the number of forprofit institutions, and a slight decline in the number of notforprofit institutions in the past decade.

Definitions of Institution Type and Institutional Control:	
Public 4year and above	Public institutions that have at least onedegree program that confers 4year degrees; yet not all degrees conferred are 4year degrees
Notforprofit 4 year and above	Notforprofit institutions that have at least onedegree program that confers 4year degrees; yet not all degrees conferred are 4year degrees
Forprofit 4year and above	Forprofit institutions that have at least onedegree program that confers 4yedegrees; yet not all degrees conferred are 4year degrees
Public 2year	Public institutions that confer 2year degrees; Often referred to as CommunityColleges
Notforprofit 2year	Notforprofit institutions that confer 2year degrees
Forprofit 2year	Forprofit institutions that confer 2year degrees

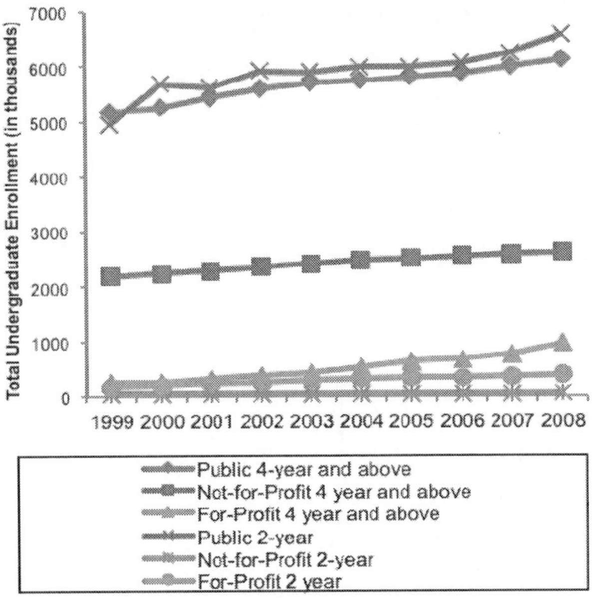

Derived from: IPEDS, WebCASPAR.

Figure C-3. Total Undergraduate Enrollments in all Fields by Institutional Sector, 1999 to 2008.

Key Points (Figure C-3 and Figure C-4):

- Fouryear and above notforprofit institutions outnumber the other institutions sectors.
- Enrollments at all institution types have been increasing steadily over the past decade.
- Public 4year and 2year institutions have larger enrollments than the other institution types.

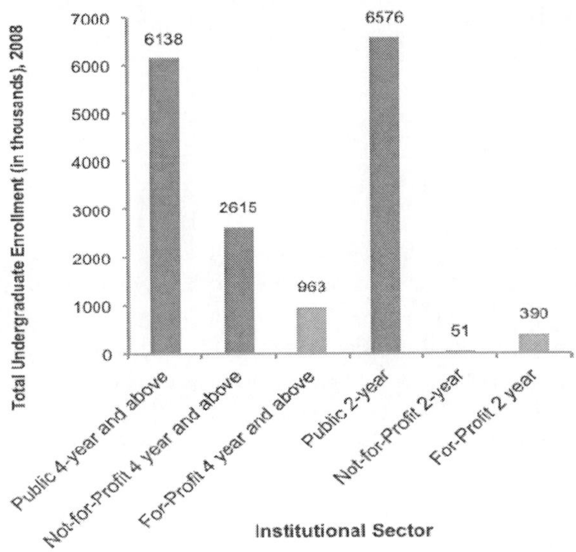

Derived from: IPEDS, WebCASPAR.

Figure C-4. Total Undergraduate Enrollments Breakdown (in thousands), 2008.

Table C-1. Distribution of Gender by Institutional Sector When First Enrolled, 2003-04

Gender			
Estimates (%)	Male	Female	Total
Total	42.5	57.5	100
First institution sector (level and control) 2003-04			
Public 4year	44.8	55.2	100
Notforprofit 4year	43.9	56.1	100
Forprofit 4year	40.3	59.7	100
Public 2year	43.4	56.6	100
Notforprofit 2year	42.5	57.5	100
Forprofit 2year	44.6	55.4	100

Derived from: U.S. Department of Education, National Center for Education Statistics, 200304 Beginning Postsecondary Students Longitudinal Study, Second Followup (BPS:04/09).

Key Points (Table C-1):

- The ratio of male students to female students across all institution sectors is fairly equal, with females representing about 5560% of the student population.

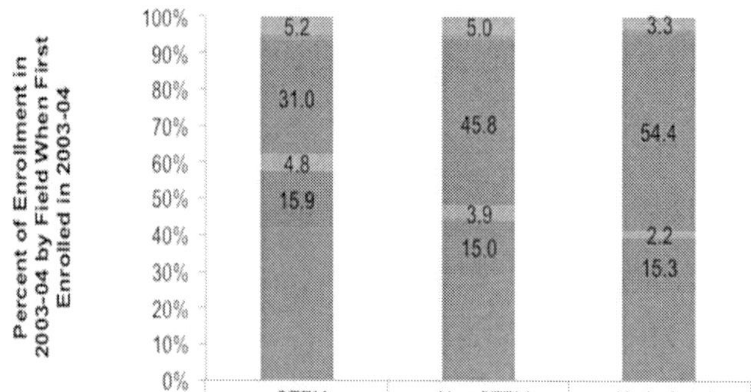

	STEM	Non-STEM	Undeclared
For-Profit 2-year	5.2	5.0	3.3
Not-For-Profit 2-year	1.4	1.1	0.8
Public 2-year	31.0	45.8	54.4
For-Profit 4-year	4.8	3.9	2.2
Not-for-Profit 4-year	15.9	15.0	15.3
Public 4-year	41.6	29.2	24.0

Field

Derived from: U.S. Department of Education, National Center for Education Statistics, 2003-04 Beginning Postsecondary Students Longitudinal Study, Second Follow-up (BPS:04/09).

Notes: For-Profit 4-year Health Professions, Not-for-Profit 2-year STEM and Health Professions, For-Profit 2-year STEM and Health Professions data should be interpreted with caution.

Estimate is unstable because the standard error represents more than 30 percent of the estimate.

Figure C-5. Percentage of Institutional Sector Enrollments by Field, 2003-04.

Key Points (Figure C-5):

- Of all students enrolled in STEM fields, 41.6% are enrolled at public 4-year institutions, 31.0% are enrolled in public 2-year institutions and 14.9% are enrolled in not-for-profit institutions.

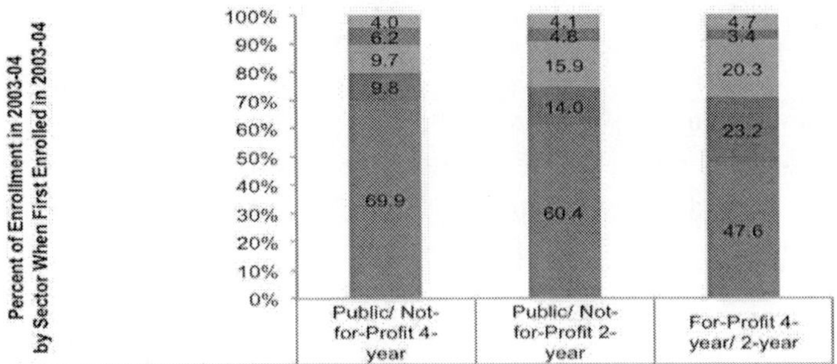

	Public/ Not-for-Profit 4-year	Public/ Not-for-Profit 2-year	For-Profit 4-year/ 2-year
Other/ More than One Race	4.0	4.1	4.7
American Indian/ Alaska Native	0.4	0.7	0.9
Asian/ Pacific Islander	6.2	4.8	3.4
Hispanic/ Latino	9.7	15.9	20.3
Black/ African American	9.8	14.0	23.2
White	69.9	60.4	47.6

Institutional Sector/Control

Derived from: U.S. Department of Education, National Center for Education Statistics, 2003-04 Beginning Postsecondary Students Longitudinal Study, Second Follow-up (BPS:04/09).

Notes: Asian/ Pacific Islander For-Profit 4-year/2-year data should be interpreted with caution. Estimate is unstable because the standard error represents more than 30 percent of the estimate. American Indian/ Alaska Native For-Profit 4-year/2-year data should be interpreted with caution. Estimate is unstable because the standard error represents more than 50 percent of the estimate.

Figure C-6. Distribution of Race/Ethnicity by Institutional Sector When First Enrolled, 2003-04.

Key Points (Figure C-6):

- Of those first enrolled across all disciplines at 4-year public or not-for profit institutions, 69.9% are white, 9.8% are black/African American and 9.7% are Hispanic/Latino.
- Of those first enrolled at the 2-year institutions, the percentage of whites is lower, and the percentages of underrepresented minorities are higher.

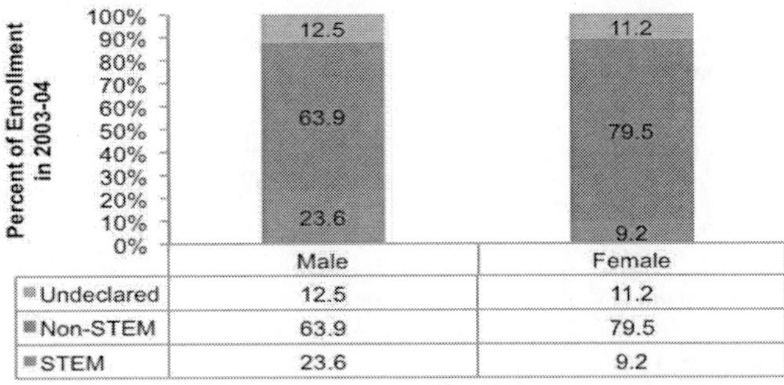

	Male	Female
Undeclared	12.5	11.2
Non-STEM	63.9	79.5
STEM	23.6	9.2

Gender

Derived from: U.S. Department of Education, National Center for Education Statistics, 2003-04 Beginning Postsecondary Students Longitudinal Study, Second Follow-up (BPS:04/09).

Notes: These data represent the type of field in which a student was last enrolled during the longitudinal survey follow up in 2009. This includes students who may have dropped out as it represents "field when last enrolled"—not taking into account whether or not the student completed a degree, was still enrolled or dropped out.

Figure C-7. Estimates of Enrolled Field by Gender.

Key Points (Table C-7):

- During the second follow-up in 2009, 23.6% of male students when last enrolled at postsecondary institutions were enrolled in STEM fields.
- During the second follow-up in 2009, 9.2% of female students when last enrolled at postsecondary institutions were enrolled in STEM fields.

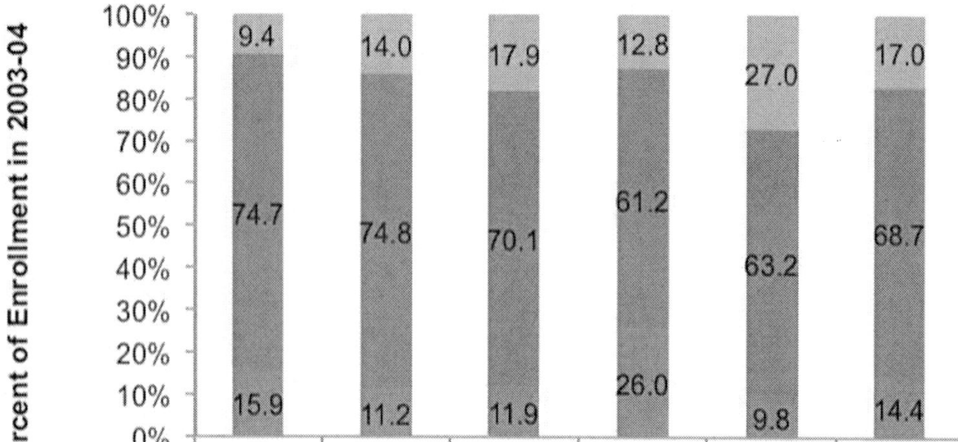

	White	Black/ African American	Hispanic/ Latino	Asian/ Pacific Islander	American Indian/ Alaska Native	Other/ More than One Race
Undeclared	9.4	14.0	17.9	12.8	27.0	17.0
Non-STEM	74.7	74.8	70.1	61.2	63.2	68.7
STEM	15.9	11.2	11.9	26.0	9.8	14.4

Race/Ethnicity

Derived from: U.S. Department of Education, National Center for Education Statistics, 2003-04 Beginning Postsecondary Students Longitudinal Study, Second Follow-up (BPS:04/09).

Notes: American Indian/ Alaska Native STEM and Health Professions data should be interpreted with caution.

Estimate is unstable because the standard error represents more than 30 percent of the estimate.

Figure C-8. Estimates of Enrolled Field by Race/Ethnicity.

Key Points (Figure C-8):

- During the second follow-up in 2009, 26.0% of Asians and 15.9% of Whites were last enrolled in a STEM field, while 9.8-11.9% of underrepresented minority groups were last enrolled in a STEM field.

Major When First Enrolled in 2003-04												
	STEM Majors											
Estimates (%)	STEM Total	Life Sci	Phys Sci	Math	Comp/ Info Sci	Engineering/ Eng Tech	Science Tech	Health Professions	Social Sci	Other Non-STEM	Undeclared	Total
Total	12.4	3.0	0.7	0.4	3.8	4.4	0.1	13.4	4.7	37.2	32.3	100
Gender												
Male	20.6	3.3	0.8	0.3	6.9	9.1	0.2 !	4.5	4.2	36.2	34.5	100
Female	6.3	2.7	0.6	0.4	1.5	1.0	0.1 !	19.9	5.1	38.0	30.7	100

Derived from: U.S. Department of Education, National Center for Education Statistics, 200304 Beginning Postsecondary Students Longitudinal Study, Second Followup (BPS:04/09).

Notes: Interpret data with caution. Estimate is unstable because the standard error represents more than 30 percent of the estimate. Interpret data with caution. Estimate is unstable because the standard error represents more than 50 percent of the estimate. These data represent the percentage of students majoring in various disciplines when first enrolled at postsecondary institutions. However, these data do not represent the fields in which these students received degrees.

Key Points (Table C-2):

- Of those enrolled at postsecondary institutions, 12.4% of all students first enroll in a STEM discipline.
- Of males enrolled at postsecondary institutions, 20.6% first enroll in a STEM discipline.
- Of females enrolled at postsecondary institutions, 6.3% first enroll in a STEM discipline.

Table C-3. Persistence in Enrolled Majors Between 2003-04 and 2009 Among Bachelor's Degree Attainers

Major When Last Enrolled 2009											
	STEM										
Estimates (%)	Life Sci	Phys Sci	Math	Comp/ Info Sci	Engineering/ Eng Tech	Sci Tech	Social Sci	Health Professions	Other Non-STEM	Undeclared	Total
Total	7.9	1.8	1.3	2.5	6	0.0 !!	17.3	6.5	56.5	0.2 !!	100
Major when first enrolled in 2003-04											
Life Sci	56.5	4.3 !	0.8 !!	0	1.6 !	0	10.5	5.4	20.9	0	100
Phys Sci	23.1	35.4	3.6 !!	0	3.8 !!	0	13.5 !	2.6 !!	18.0 !	0	100
Math	2.3 !!	2.3!!	54.9	1.4 !!	4.3 !!	0	15.3 !	0	19.5 !	0	100
Comp/Info Sci	1.2 !!	0	2.6 !	48.1	4.3 !	0	11.4	0	32	0.4 !!	100
Engineering/ Eng Tech	4.6 !	1.3 !	2.2 !	4.2 !	65.1	0	3.3 !	0.7 !!	17.8	0.9 !!	100
Social Sci	2.5 !	0.3 !!	0.3	0.1 !!	1.3 !!	0	65.7	1.2 !!	28.6	0	100
Sci Tech	‡	‡	‡	‡	‡	‡	‡	‡	‡	‡	100
Health Professions	13.8	1.2 !	0	1.1 !!	0.7 !!	0	7.8	43	32.4	0	100
Other NonSTEM	1.8	0.3 !	0.9 !	1	0.9	0	9.3	2.9	82.9	0	100
Undeclared	7.8	2.6	0.7 !	1.9 !	3.1	0.0 !!	22.7	5.1	55.5	0.5 !!	100

Derived from: U.S. Department of Education, National Center for Education Statistics, 200304 Beginning Postsecondary Students Longitudinal Study, Second Followup (BPS:04/09).

Notes: ‡ Reporting standards not met. ! Interpret data with caution. Estimate is unstable because the standard error represents more than 30 percent of the estimate. !! Interpret data with caution. Estimate is unstable because the standard error represents more than 50 percent of the estimate.

Key Points (Table C-3):

- Among students first enrolled in Life Sciences in 200304, 56.5% attained a bachelor's degree in Life Sciences by 2009.
- Among STEM fields, bachelor's degree attainment within the field of original enrollment was highest among engineering and engineering technician enrollees at 65.1%.

	Attainment or level at last institution enrolled through 2009						
	Completers			Persisters		Leavers	
Estimates (%)	Attained bachelor's degree	Attained associate's degree	Attained certificate	No degree, enrolled at 4-year	No degree, enrolled at lessthan-4-year	No degree, not enrolled	Total
Total	30.7	9.3	9.4	7.1	7.9	35.5	100
Field when First Enrolled in 2003-04							
STEM	40.8	10.0	3.7	8.6	5.5	31.5	100
Non-STEM	30.0	10.2	10.6	6.8	7.6	34.8	100
Undeclared	28.1	7.6	9.7	6.9	9.5	38.2	100

Derived from: U.S. Department of Education, National Center for Education Statistics, 2003-04 Beginning Postsecondary Students Longitudinal Study, Second Follow-up (BPS:04/09).

Key Points (Table C-4):

- Students first enrolled in STEM fields have a higher degree of postsecondary degree attainment than students first enrolled in other fields.
- 40.8% of students first enrolled in a STEM field attained a bachelor's degree in any field, while 30.0% of students in a non-STEM field and 28.8% of students who were first undeclared attained a bachelor's degree.

Table C-5. 2009 Attainment Level by Major When First Enrolled in 2003-04

Attainment or level at last institution enrolled through 2009							
	Completers		Persisters		Leavers		
Estimates (%)	Attained bachelor's degree	Attained associate's degree	Attained certificate	No degree, enrolled at 4-year	No degree, enrolled at less-than-4-year	No degree, not enrolled	Total
Total	30.7	9.3	9.4	7.1	7.9	35.5	100
Major when first enrolled in 2003-04							
Life Sci	56.7	8.6	1.8 !	9.3	4.0	19.5	100
Physical Sci	57.9	6.1 !!	3.7 !	8.7 !	4.6 !	18.9	100
Math	54.6	7.4 !	4.1 !	0.0	1.7 !!	32.3	100
Comp/ Info Sci	20.6	15.6	3.8	8.6	7.4	44.0	100
Engineering/ Eng Tech	44.6	8.0	4.6 !	8.8	5.7	28.3	100
Science Tech	35.9 !	11.8 !!	1.8 !!	4.3 !!	19.4 !!	26.9 !	100
Social Sci	50.7	6.7	1.4	8.0	8.6	24.5	100
Health Professions	17.7	11.6	19.3	4.8	9.7	37.0	100
Other NonSTEM	31.7	10.1	8.7	7.4	6.6	35.6	100

Derived from: U.S. Department of Education, National Center for Education Statistics, 2003-04 Beginning Postsecondary Students Longitudinal Study, Second Follow-up (BPS:04/09).

Notes: ! Interpret data with caution. Estimate is unstable because the standard error represents more than 30 percent of the estimate. !! Interpret data with caution. Estimate is unstable because the standard error represents more than 50 percent of the estimate.

Key Points (Table C-5):

- Students first enrolled in life sciences or physical sciences are more likely to attain bachelor's degrees in **any** field (not necessarily a STEM field) than students first enrolled in non-STEM fields.

	Degree Attainment and Persistence as of 2009					
	Completers		Persisters		Leavers	
Estimates (%)	Attained a STEM degree	Attained non-STEM degree	No degree; enrolled in a STEM field	No degree; enrolled in a non-STEM field	Left postsecondary education without a degree	Total
Total	8.1	42.2	1.8	13.5	34.5	100
Major field of study in 2003-04						
STEM	35.1	21.6	5.7	8.9	28.7	100
• Math	40.3	27.3	0.0	1.9 !!	30.5 !	100
• Life Sci	37.8	31.9	3.7	9.7	16.8	100
• Physical Sci	41.3	28.1	2.1 !!	11.5 !	16.9	100
• Eng/ Eng Tech	41.8	16.9	7.2	7.9	26.2	100
• Comp/ Info Sci	24.6	16.7	6.6 !	9.3	42.7	100
• Science Tech	‡	‡	‡	‡	‡	100
NonSTEM	3.1	48.7	1.0	13.6	33.6	100
Undeclared	6.1	39.1	1.6	15.0	38.3	100

Derived from: Preliminary Estimate - U.S. Department of Education, National Center for Education Statistics, 2003-04 Beginning Postsecondary Students Longitudinal Study, Second Follow-up (BPS:04/09).

Key Points (Table C-6):

- Among students first enrolled in STEM fields, 35.1% of students attained a STEM degree.
- Among the STEM fields, students first enrolled in physical science and engineering/engineering technology had the highest percentage of degree attainment within a STEM field at above 40% while computer science had the lowest at 24.6%.

Table C-7. STEM and Non-STEM Attainment by Demographic Characteristics

	Persistence in STEM field (excl social/behavioral sci) as of 2009					
	Completers		Persisters		Leavers	
Estimates (%)	Attained a STEM degree	Attained non-STEM degree	No degree; enrolled in a STEM field	No degree; enrolled in a non-STEM field	Left without a degree	Total
Total	8.1	42.2	1.8	13.5	34.5	100
Gender						
Male	12.1	35.3	3.2	13.2	36.1	100
Female	5.1	47.3	0.7	13.6	33.3	100
Race/ethnicity						
White	9.3	46.2	1.8	10.9	31.9	100
Black/ African American	4.1	32.6	1.2	19.5	42.7	100
Hispanic/ Latino	4.8	36.9	1.6	15.3	41.3	100
Asian	15.9	40.6	3.7 !	17.6	22.2	100
American Indian/ Alaska Native	5.7 !!	25.5	0.3 !!	26.0 !	42.6	100
Native Hawaiian/ other Pacific Islander	4.9 !!	36.9 !	0	22.7 !!	35.5 !	100
Other	7.9 !	37.8	1.4 !!	21.1	31.7	100
More than one race	7.0	37.5	2.7 !	16.8	36	100

Derived from: Preliminary Estimate - U.S. Department of Education, National Center for Education Statistics, 2003-04 Beginning Postsecondary Students Longitudinal Study, Second Follow-up (BPS:04/09).

Key Points (Table C-7):

- Overall, 8.1% of who entered postsecondary education in 2004/5 had attained a STEM degree by 2009.
- Among male and female students who entered postsecondary education in 2004/5, 12.1% and 5.1% attained a STEM degree, respectively.
- Asian and White students had the highest percent of degree attainment within a STEM field at approximately 16% and 9%, respectively, while a distribution of 4-6% of underrepresented minorities attained STEM degrees within their race/ethnicity groups.

Percentage of Degrees Conferred in 2009											
STEM Majors											
Race/Ethnicity	STEM Total	Life Sci	Physical Sci	Math	Comp/ Info Sci	Engineering/ Eng Tech	Sci Tech	Social Sci	Health Professions	Other Non-STEM	Total
White	14.4	4.9	1.1	0.7	2.7	5.0	0.1	9.7	12.3	63.5	100
Black	10.6	2.9	0.6	0.4	3.4	3.3	0.1	10.2	13.9	65.3	100
Hispanic	11.9	3.4	0.7	0.5	2.6	4.7	0.1	10.5	10.8	66.8	100
Asian/ Pacific Islander	23.0	10.0	1.7	1.1	3.0	7.1	0.1	12.5	11.0	53.5	100
American Indian/ Alaska Native	13.1	4.9	1.0	0.4	2.6	4.1	0.1	9.6	12.6	64.7	100

Derived from: NCES, IPEDS, WebCASPAR.

Key Points (Table C-8):

- Among all White students who received degrees in 2009, 14.4% obtained degrees in a STEM field.
- Across all race/ ethnicity categories, White and Asian/Pacific Islander students obtained STEM degrees at the highest percentages at 14% and 23%, respectively. That is, among the Asians who earned degrees in 2009, 23.0% of the degrees are in STEM. Among underrepresented minorities, 10.6% of Blacks and 11.9% of Hispanics earned degrees in STEM.

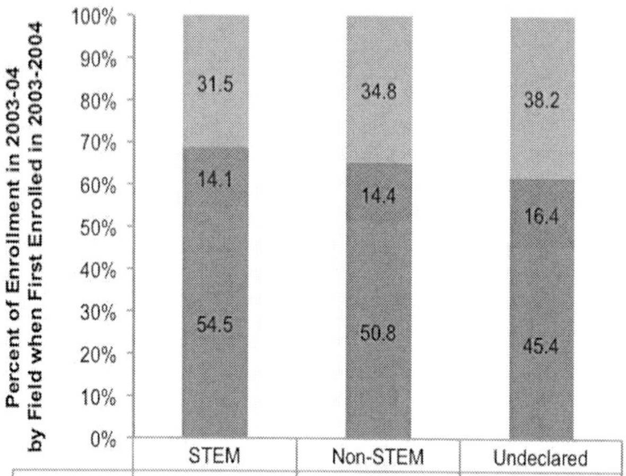

	STEM	Non-STEM	Undeclared
Leavers	31.5	34.8	38.2
Persisters	14.1	14.4	16.4
Completers	54.5	50.8	45.4

Field

Derived from: U.S. Department of Education, National Center for Education Statistics, 2003-04 Beginning Postsecondary Students Longitudinal Study, Second Follow-up (BPS:04/09).

Notes: The attainment levels are aggregated into three higher-level categories: completers, persisters, and leavers.

Figure C-9. 2009 Attainment Level by Discipline When First Enrolled in 2009.

Key Points (Figure C-9):

- Overall, 31.5% of students first enrolled in STEM had no degree and were no longer enrolled after the 6 year follow-up, while 34.8% of non-STEM students and 38.2% of undeclared students had dropped out.

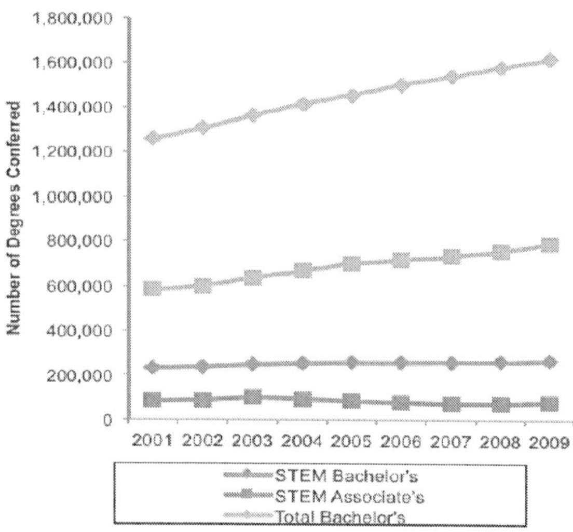

Derived from: NCES, IPEDS, WebCASPAR.

Figure C-10. STEM and Total Bachelor's and Associate's Degrees Conferred, 2001-2009.

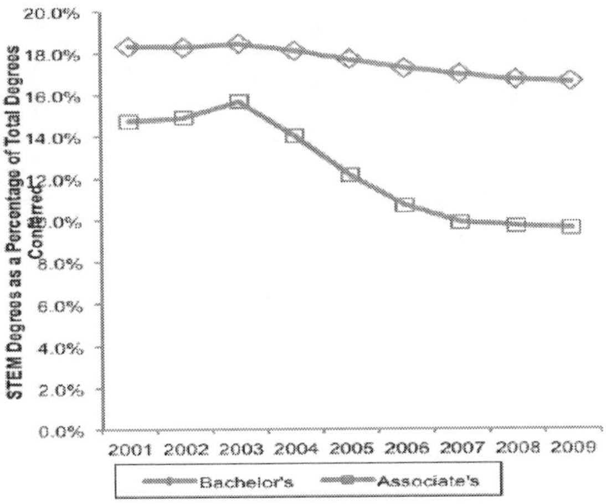

Derived from: NCES, IPEDS, WebCASPAR.

Figure C-11. STEM Bachelor's and Associate's Degrees as a Percentage of Total Degrees Conferred, 2001-2009.

Key Points (Figure C-10 and Figure C-11):

- A greater number of bachelor's degrees are conferred than associate's degrees.
- A greater percentage of bachelor's degrees than associate's degrees are conferred in STEM fields.
- STEM degrees as a percentage of all degrees conferred has declined since 2001 at both the bachelor's and associate's degree levels.
- The percentage of STEM associate's degrees conferred decreased sharply between 2003 and 2005, but has since leveled off. This trend is mostly due to the rise in associate's degrees in health professions and the decrease in associate's degrees in computer/information sciences.

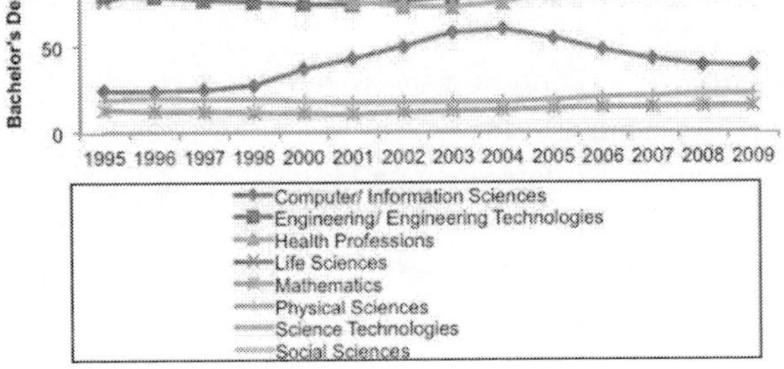

Derived from: NCES, IPEDS, WebCASPAR.

Figure C-12. STEM, Health, and Social Science Bachelor's Degrees Conferred, 1995-2009.

Key Points (Figure C-12):

- Computer/information science bachelor's degrees increased from 2000 to 2004, but reverted back to 2000 levels from 2005 to 2009.
- Engineering/engineering technologies bachelor's degrees have remained steady over the past decade.

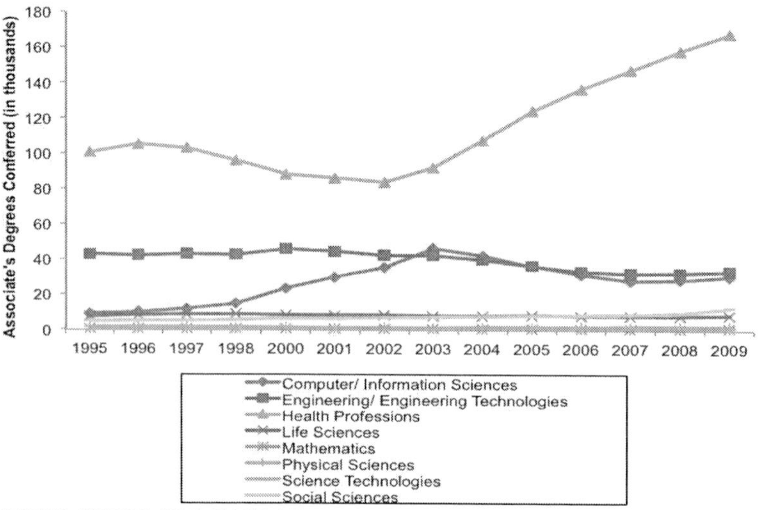

Derived from: NCES, IPEDS, WebCASPAR.

Figure C-13. STEM, Health, and Social Science Associate's Degrees Conferred, 1995-2009.

Key Points (Figure C-13):

- Compared to conferred bachelor's degrees, a greater proportion of associate's degrees are conferred in health professions and computer/ information sciences.

Appendix D: Economic Analysis of STEM Workforce Need

Appendix D describes data regarding the demand for STEM workers and the anticipated supply of STEM undergraduates from post-secondary higher education institutions. This appendix is the result of work done by IDA's Science and Technology Policy Institute.

- *STEM Professionals.* Workers who regularly draw on their expertise in STEM fields—including scientists, engineers, mathematicians, and technicians in STEM occupations—make up a relatively small but essential fraction of the U.S. workforce. They advance the frontiers of knowledge in industry, government, and academia, generating the new ideas and technologies that can transform entire industries and sectors of society. In colleges and universities, they also educate future generations of scientists, engineers, technicians, and mathematicians along with other students who will draw on STEM knowledge throughout their lives.

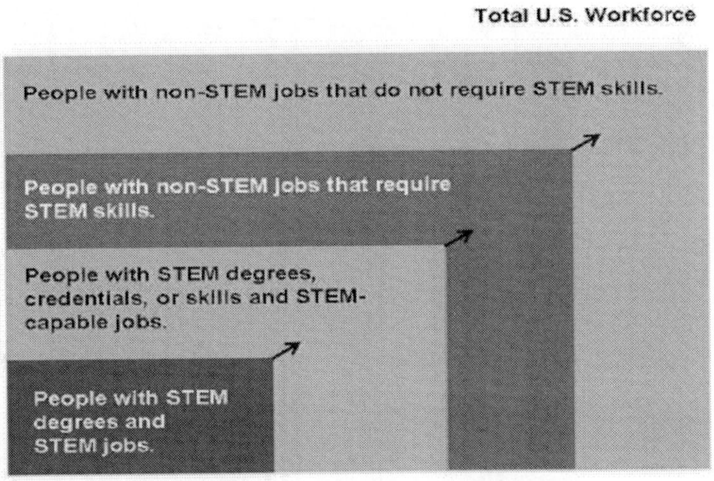

Note: The categories of jobs that require STEM skills and understandings are expanding, generating additional demand for workers with STEM degrees.

Figure D-1. STEM Workforce Definition.

- *The STEM-Capable Workforce*. A much larger group of workers, whom we categorize in this report as the STEM-capable workforce, routinely use knowledge and skills developed in STEM fields as part of their jobs. Many of these people have STEM degrees or certificates but are working in jobs that would not be formally categorized as STEM occupations. At one end this group shades into the ranks of STEM professionals who develop and apply new knowledge. At the other end it shades into workers in all professions who use information and capabilities derived from science, technology, engineering, and mathematics to analyze, communicate, innovate, manage, and strategize. For example, physicians, nurses and other health workers generally are not categorized as STEM professionals, yet many of these individuals draw heavily on STEM knowledge and skills in their jobs. As another example, the advanced manufacturing workforce requires proficiency in math, technology, and engineering principles to succeed in their jobs, from entry-level workers through graduate-degreed engineers.
- *Non-STEM Workers Who Draw on STEM Skills*. Many occupations today require higher levels of familiarity with STEM subjects than they did in the past.[121] A proxy for these increased demands is the increasing level of education required for many jobs. Between 1973 and 2008, the share of jobs in the U.S. economy that required postsecondary education increased from 28% to 59%, and this percentage is projected to continue to increase.[122] While college provides knowledge and skills other than STEM capabilities, the prominence of STEM subjects in higher education suggests that at least part of what employers are seeking is greater familiarity with STEM concepts and skills and STEM-derived technologies. One of many examples would be a market researcher who uses statistical techniques to draw conclusions; such a worker might fall into either this category or the previous category depending on the exact nature of the job.

Table D-1. Estimated Number of Employed Persons and Percentage of Unemployed (Compared to Entire Labor Force), 2005-2009

	2005		2006		2007		2008		2009	
Occupation	Employed	% Unemp.*	Employed	% Unemp.*	Employed	% Unemp.*	Employed	% Unemp.*	Employed	% Unemp.*
All Occupations	137,147,492	6.9%	142,545,193	6.4%	143,630,939	6.3%	147,486,711	6.3%	141,932,926	9.8%
All STEM Occupations	7,112,960	3.3%	7,347,542	2.6%	7,369,304	2.5%	7744152	2.6%	7,498,307	5.2%
All STEM Managers	548,618	2.9%	581,833	2.0%	602,188	1.7%	645,681	2.4%	635,896	4.6%
All Computer Occupations	2,996,823	4.0%	3,110,616	2.9%	3,164,358	2.9%	3,297,403	2.8%	3,315,697	5.2%
All Mathematical Occupations	162,123	2.1%	163,212	2.1%	168,721	2.6%	178,357	1.5%	187,337	3.4%
All Engineers	1,833,225	2.6%	1,854,730	1.9%	1,837,649	1.7%	1,906,983	1.9%	1,777,445	4.4%
All Engineering Technicians	686,259	4.0%	710,790	4.0%	728,494	3.4%	767,355	4.0%	650,708	9.5%
All Life and Physical Scientists	382,472	2.7%	394,300	2.3%	423,308	2.2%	430,841	2.3%	625,330	3.0%
All Life Scientists	232,008	1.5%	254,607	1.3%	239,823	2.0%	285,454	2.0%	266,868	2.0%
All Physical Scientists	346,370	2.5%	357,387	1.8%	326,480	2.4%	360,300	2.3%	358,462	3.8%
All Science Technicians	307,534	4.4%	314,367	4.5%	301,591	4.6%	302,619	4.6%	305,894	6.9%

Derived from: American Community Survey oneyear estimates, Census Bureau; data retrieved from IPUMSUSA database.

Note: * %Unemp. indicates percentage of labor force that is unemployed.

Key Points (Table D-1):

- STEM occupations make up approximately 5.6% of the employed labor force in the United States.
- From 2005 through 2009, STEM occupations fared well compared to the general workforce; the total unemployment percentage for all STEM jobs was typically about half of that for the entire occupational labor force.
- Overall, the number of STEM workers in the labor force grew from 7.5 million in 2005 to over 7.9 million in 2009

- *Non-STEM Workers Who Do Not Draw on STEM Skills.* Many jobs in the economy do not draw directly on STEM skills. To again cite a specific example, athletes, singers, actors, and other entertainers typically do not draw on STEM subjects to do their jobs. However, even these individuals may need to master specific STEM content—for example, to devise a training regimen, or to create or disseminate artistic materials using new technologies.

In general, no job is completely isolated from the influence of new technologies and new ideas. All Americans regularly encounter the products of science, technology, engineering, and mathematics in their jobs and in their daily lives, though they may not recognize the connection with STEM subjects.

The decisions individuals make in supermarkets, doctors' offices, and voting booths often depend at least in part on ideas drawn from STEM fields. To the extent that people are comfortable and familiar with STEM concepts, they are better able to take advantage of new opportunities and make good decisions on STEM-related issues. In doing so, they help create a cultural environment that is conducive to STEM endeavors and to the benefits those endeavors can produce.

Table D-2. Fastest Growing Jobs as Reported by the Bureau of Labor Statistics, 2008–2018

Occupation	Percent Change	Number of New Jobs	Median Annual Wage (May 2008)
Biomedical Engineers	72	11,600	$77,400
Network Systems and Data Communications Analysts	53	155,800	$71,100
Home Health Aides	50	460,900	$20,460
Personal and Home Care Aides	46	375,800	$19,180
Financial Examiners	41	11,100	$70,930
Medical Scientists (Except Epidemiologists)	40	44,200	$72,590
Physicians Assistants	39	29,200	$81,230
Skin Care Specialists	38	14,700	$28,730
Biochemists and Biophysicists	37	8,700	$82,840
Athletic Trainers	37	6,000	$39,640

Derived from: Bureau of Labor Statistics Occupational Employment Statistics and Division of Occupational Outlook.

Note: Highlighted rows indicate STEM-related occupations.

Key Points (Table D-2):

- Though the Bureau of Labor Statistics does not have an official STEM designation for categorizing occupations, those commonly labeled as STEM in other research appear as some of the fastest growing in the most recent employment projections.
- While the absolute number of new jobs being created for biomedical engineers as well as biochemists and biophysicists remains relatively low compared to others on this list, the median wages earned by all of the fastest growing STEM occupations are some of the highest among all occupations.

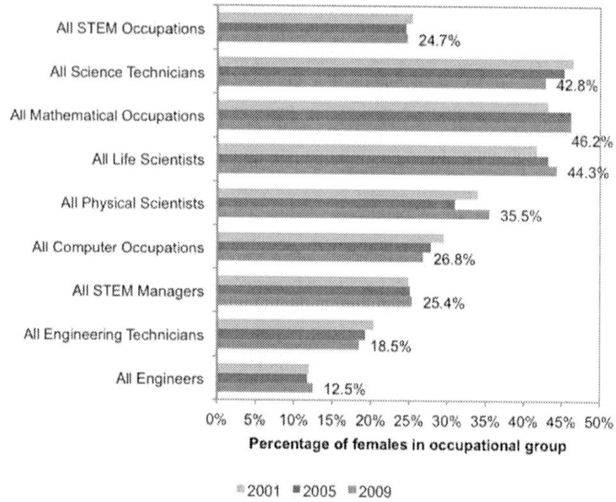

Derived from: American Community Survey of The Census Bureau, 2001, 2005, 2009 (one-year estimates); data retrieved from IPUMS-USA database.

Figure D-2. Estimated Percentages of Females in STEM Occupational Groups, 2001, 2005, and 2009 (Data Labels Indicate 2009 Values).

Key Points (Figure D-2):

- Overall, the average percentage of females in all STEM occupations in 2009 (25%) was the same as in 2001.
- The ratio of female to male workers remains low despite large numbers of women entering selected occupational fields (e.g., life sciences) over the past decade.
- The number of women in STEM occupations in 2009 ranged from as low as 9% in some engineering fields to upwards of 46% in mathematical occupations.

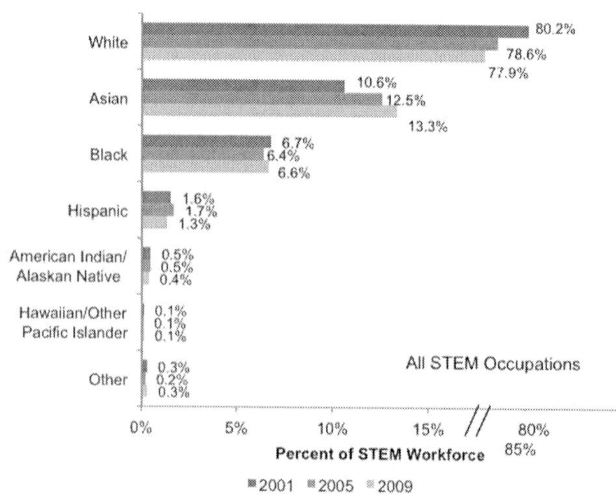

Derived from: American Community Survey of The Census Bureau, 2001, 2005, 2009 (one-year estimates); data retrieved from IPUMS-USA database.

Note: Detailed data for all ethnicities and STEM occupational groups are available in Appendix C.

Figure D-3. Estimated Race/Ethnicity of Labor Force in All STEM Occupations, 2001, 2005, and 2009.

Key Points (Figure D-3):

- The percentage of various race/ethnicities across STEM occupations has remained stable from 2001 through 2009. One exception is Asians, who have moved from 10.6% of all STEM occupations in 2001 to 13.7% in 2009.
- The trends in race/ethnicity vary when looking at specific occupational groups. For example, in the life sciences (not shown), the percentage of Whites has decreased significantly from 2001 to 2009, and much of that employment has shifted to Asians.

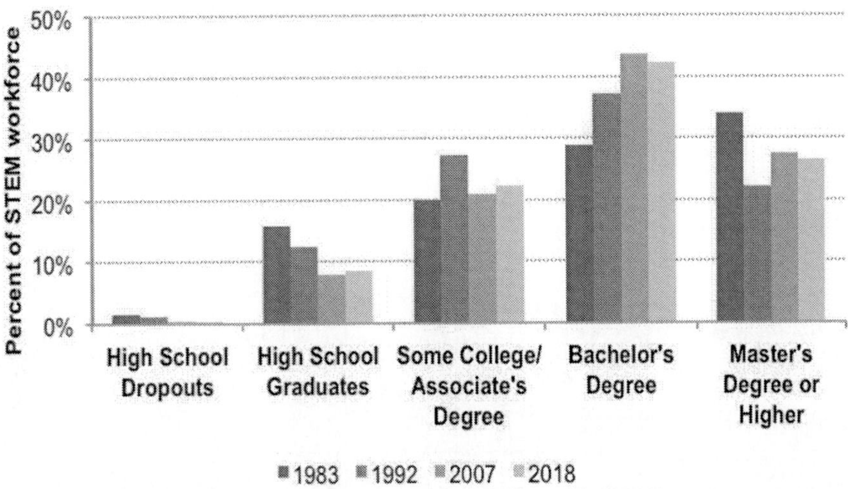

Derived from: Carnevale, Anthony, Nichole Smith, and Jeff Strohl. (2010). *Help Wanted: Projections of Jobs and Education Requirements through 2018*. Washington, DC: Georgetown University Center on Education and the Workforce.

Note: The definition of STEM used here includes the following occupations and corresponding standard occupational classifications (SOC): computer and mathematical science occupations (SOC 15-1011-SOC 15-2099), architecture and architectural technician occupations (SOC 17-1011-SOC 17-1022; SOC 17-3012-SOC 17-3019; SOC 17-3031); engineers and engineering technician occupations (SOC 17-2011-SOC 17-2199; SOC 17-3021-SOC17-3031); life and physical sciences occupations (SOC 19-1011-SOC19-2099; SOC 19-4011- SOC 19-4099); social sciences occupations (SOC 19-3011-SOC19-3099).

Figure D-4. Historical and Projected Educational Attainment in STEM Occupations, Various Years from 1983 through 2018.

Key Points (Figure D-4):

- Based on the Center on Education and the Workforce's projections and a historical analysis of BLS data, the authors revealed a trend through 2018 that more STEM occupations will demand education that includes at least some college.
- Over time, the population of STEM workers that are high school dropouts or high school graduates decreases to 8.8% by 2018, thus indicating that 91.2% of STEM workers will need at least some post-secondary education.

Table D-3. Percent of Labor Force with Bachelor's STEM Degrees (Columns) in Corresponding STEM Occupations (Rows)

	STEM Bachelor's Degrees											
	All Comp & Info Sys	All Mathematics	All Engineering	All Eng Techs	All Life Sciences	All Phys Sciences	All Science Techs	All STEM Education	Eng and Industrial Mgmt	Multidisciplinary or General Science	% of Corresponding Labor Force with any STEMBachelor's Degree	All Comp & Info Sys
STEM Occupations	All Computer Occupations	18.8%	3.1%	12.3%	0.9%	1.7%	2.0%	0.1%	0.2%	0.1%	0.7%	40.0%
	All Mathematical Occupations	5.9%	16.0%	6.0%	0.4%	3.4%	2.2%	0.1%	0.5%	0.1%	0.8%	35.5%
	All Engineers	1.2%	0.7%	43.1%	2.0%	1.4%	1.7%	0.2%	0.1%	0.1%	0.5%	50.9%
	All Engineering Technicians	0.5%	0.2%	4.9%	1.2%	1.5%	0.7%	0.1%	0.0%	0.0%	0.2%	9.4%
	All Life Scientists	0.5%	0.7%	4.1%	0.1%	51.7%	9.2%	0.0%	0.1%	0.1%	1.9%	68.5%
	All Physical Scientists	1.5%	1.8%	9.8%	0.3%	27.0%	35.4%	0.2%	0.5%	0.0%	2.1%	78.5%
	All Science Technicians	0.5%	0.4%	2.8%	0.2%	15.0%	5.1%	0.2%	0.3%	0.0%	0.7%	25.1%
	All STEM Managers	12.3%	2.8%	22.4%	1.4%	2.8%	2.4%	0.1%	0.2%	0.1%	0.6%	45.3%
	All STEM Occupations	9.9%	2.3%	18.8%	1.1%	5.2%	3.8%	0.1%	0.2%	0.1%	0.7%	42.3%

Derived from: American Community Survey of the Census Bureau, 2009. Data retrieved from IPUMSUSA database.

Note: This table excludes social scientists from disciplines and occupations; a complete listing of the columnar degree field groups and the corresponding detailed degree fields is detailed below; occupations follow the CPST conventions of ~50 detailed occupations aggregated into eight groups but excluding those occupations in the social sciences.

Key Points: (Table D-3):

- This table illustrates the difficulty in tracking the STEM workforce due to the number of possible occupational paths that a STEM graduate may take; only the options categorized as STEM occupations are shown here but many more STEM enabled occupations exist.

- Across all education levels, 42.3 percent of individuals in the STEM ... STEM field while 64.4 percent of individuals in the STEM labor force with a bachelor's degree or above received their bachelor's in a STEM field (data not shown).
- Life science occupations have the highest proportion of bachelor's degrees from the identical discipline of life sciences with 51.7% working in the same field they studied. The two groups with the lowest number of workers who have a matching degree and occupational field are engineering technicians and science technicians.
- Of the estimated 11.4 million individuals with STEM bachelor's degrees in the entire workforce in 2009, only 3.3 million or 29.3 percent of them were in STEM occupations.

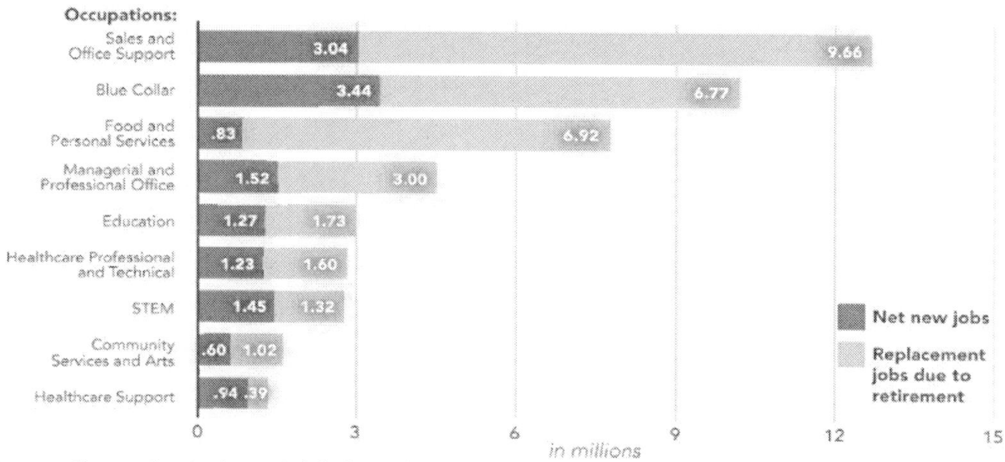

Source: Carnevale, Anthony, Nichole Smith, and Jeff Strohl. (2010). *Help Wanted: Projections of Jobs and Education Requirements through 2018*. Washington, DC: Georgetown University Center on Education and the Workforce. p. 27. Reprinted with permission.

Figure D-5. New and Replacement Job Openings and Occupational Distribution.

Key Points (Figure D-5):

- Based on Carnevale et al. (2010), net new STEM jobs and STEM replacement jobs due to retirement are projected to be about 2.77 million between 2008 and 2018.

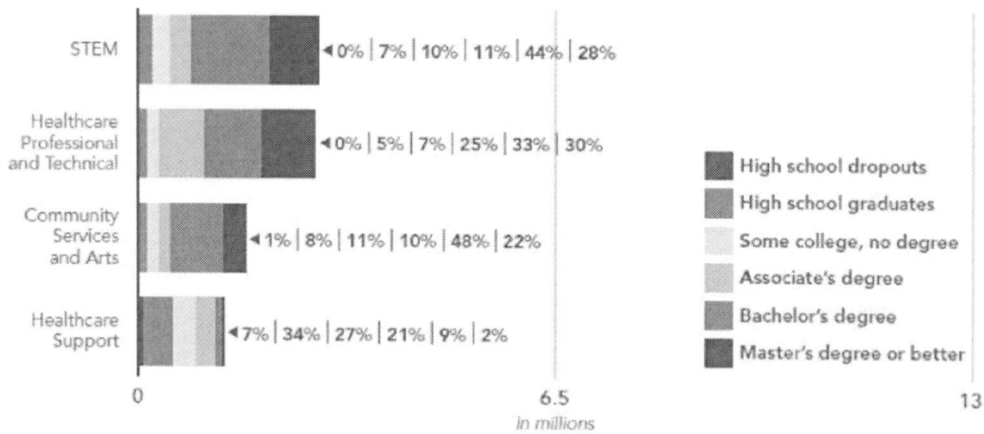

Source: Carnevale, Anthony, Nichole Smith, and Jeff Strohl. (2010). *Help Wanted: Projections of Jobs and Education Requirements through 2018*. Washington, DC: Georgetown University Center on Education and the Workforce. p. 28. Reprinted with permission.

Figure D-6. Total Job Openings and the Distribution of Educational Demand within Occupations.

Key Points (Figure D-6):

- 83% of these jobs require an associate's degree or above.
- The projected total number of new and replacement STEM jobs between 2008 and 2018 requiring an associate's degree or above is 2.3 million.

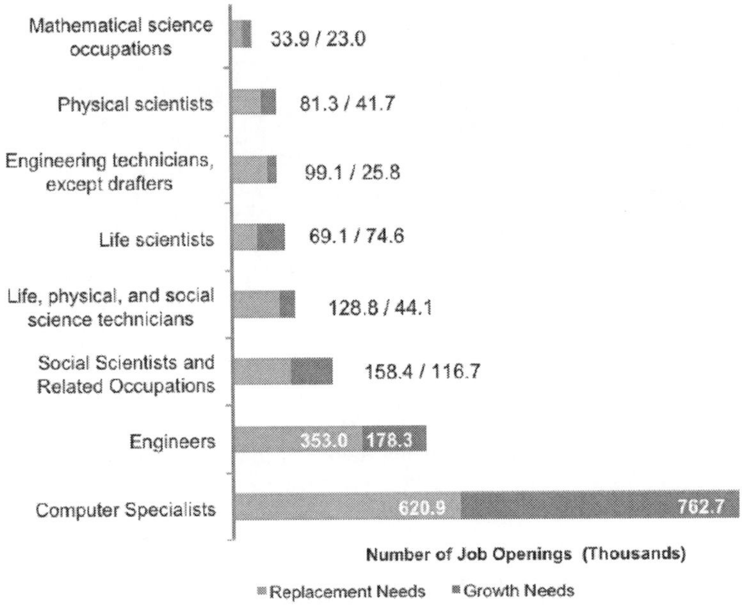

Derived from: Bureau of Labor Statistics.

Figure D-7. Projected Job Openings in STEM Occupations, 2008–2018.

Key Points (Figure D-7):

- The number of job openings projected in 2018 is delineated here as growth versus replacement needs. Job growth includes the creation of new jobs while replacement needs are those that result from workers retiring or permanently leaving a position. Together, these categories indicate the minimum number of workers who will need to be trained for the given occupation.
- In terms of total number of job openings, computer specialists are projected to require more than 1.3 million workers as a result of job growth and replacement needs.
- Engineers are the next most required STEM occupation with 531,300 job openings projected through 2018.

APPENDIX E: EVIDENCE OF THE MATHEMATICS PREPARATION GAP

Today, many students enter college not prepared for college level mathematics. Among students who take the ACT entrance examination, just 43 percent achieve the ACT College Readiness Benchmark in mathematics.[123] Because of inadequate preparation, many students need to take developmental classes in mathematics when they get to college. This poses a burden on students, institutions of higher education, the military, and employers in the form of developmental education and worker training. Higher education alone spends at least two billion dollars on developmental education per year.[124]

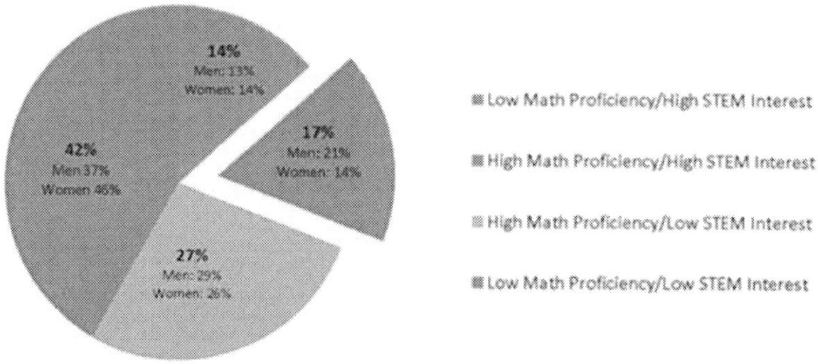

Source: The Business-Higher Education Forum. (2011). The STEM interest and proficiency challenge: Creating the workforce of the future. Washington, DC.

Figure E-1. 12th Grade Student STEM Interest and Mathematics Proficiency.

Key Points (Figure E-1)

- Among high school seniors who have taken the series of exams offered by the ACT in eighth, tenth, and twelfth grades, about one in six is both proficient in mathematics and interested in STEM fields.

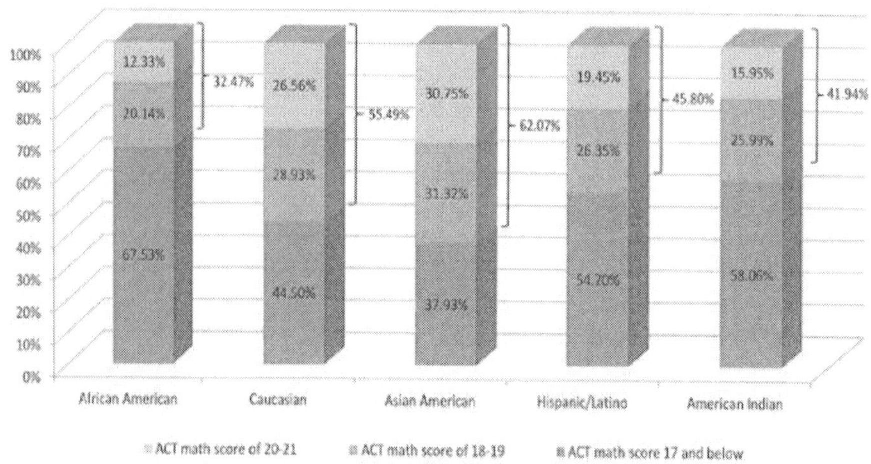

Source: ACT. (2011). The Condition of College & Career Readiness 2011. Iowa City, IA: ACT.
Note: Students who score a 22 or higher on the mathematics portion of the ACT exam are considered math proficient and have a high probability of college success.

Figure E-2. High School Student Performance on ACT Math Exam..

Key Points (Figure E-2)

- Closing the mathematics-preparation gap would enable many more students to pursue STEM degrees in college. About 15% of 12th graders are interested in STEM fields but not proficient in math, with women slightly more common in this category (Figure E-1). Furthermore, many members of this group are not far from math proficiency. More than half of white and Asian-American students, more than 40% of Hispanic/Latino and American Indian students, and almost one third of African-

American students who are interested in STEM fields are within four points on the ACT exam of the cutoff for math proficiency (Figure E-2). If the preparation of these students in math could be enhanced, many more students could be prepared to pursue STEM fields in college.

- Large numbers of students who take the ACT exam in twelfth grade, including many students from groups underrepresented in STEM fields, are within a few points on the exam of mathematical proficiency.
- One idea to improve and decrease the high cost of math remediation is to make widespread availability of the resources the Federal government has developed for its use in training the U.S. military (See, for example, Box E-1).

Box E-1: Using Artificial Intelligence to Bridge the Mathematics-Preparation Gap

The Office of Naval Research (ONR), the science and technology provider for the U.S. Navy and Marine Corps, has supported academic research in cognitive learning science for more than two decades.

For example, the Cognitive Science of Learning program has supported the development of computer-based learning tools, including a 3-D video game developed by ONR, that permit recruits to learn at-sea safety, ship handling, and electronics maintenance during on-shore training.

Recruits who use the safety video game make 50 percent fewer errors and locate ship or submarine compartments in 50 percent less time than others. In a study measuring how much information recruits remember, game-playing recruits retained 83 percent of their reading gains, almost four times more than their counterparts.[a]

ONR is now developing artificially intelligent STEM tutors to help high school students increase their proficiency in STEM subjects. ONR-sponsored researchers at Arizona State University have demonstrated the success of digital tutors among algebra students, raising student grade levels by up to 20 percent--the equivalent of increasing going from "Cs" to "As."[b]

The success of these intelligent tutors has led the Chief of Naval Research to sponsor a multi-million dollar "grand challenge" to adapt the technology for use in STEM education projects.

Sources:

[a] Murphy, C. (In Press). *Why Games Work—The Science of Learning.* Modsim World 2011, Virginia Beach, VA, October 2011. Accesible from http://www.goodgamesby design. com/?p=59.

[b] Barrus, A. K. Sabo, S. Joseph, R. Atkinson, and R. Perez. (In Press). E*valuating Adaptive, Computer-Based Mathematics Tutoring Systems: A Math Improvement Feasibilty Study.* Tempe, AZ: Arizona State University.

APPENDIX F: EFFICACY OF VARIOUS CLASSROOM METHODS

Thinking like a scientist requires acquisition of information, habits of mind, skills, and a scientific identity.[125] It seems unlikely that such diverse attributes would all be learned most effectively through one mode of teaching. Indeed they are not. Yet most introductory STEM courses taken in the first two years of college are in the same format: lectures, followed by practice problem sets, followed by multiple choice or word-problem tests. A single model of instruction cannot achieve all the significant learning goals of science instruction, nor can a single form of assessment detect all the consequential outcomes.

To create vibrant science classrooms that effectively transmit knowledge and develop the intellectual attributes of scientists, college faculty must overcome the inertia of the historical habits passed from generation to generation.

A substantial empirical literature demonstrates that alternative models of instruction can achieve many important learning outcomes more effectively than current practice and without added time or cost (for one example, see Box F-1). These studies address learning in many fields of science as we well as engineering and math.

Many of the alternatives include lectures, but they also include two key elements: (1) Students are actively engaged in the process of learning compared to solely following a lecture and then executing what they have been told; and (2) Students receive feedback while learning, which is usually inherent in activities that engage students' minds.

Two types of studies demonstrate the impact of active learning on comprehension of concepts and retention of information. The first are randomized, controlled studies conducted under experimental laboratory conditions in which students are taught the same material in different ways.

One study determined that either writing or talking about material increased comprehension and learning over a control group, and both talking and writing had a more substantial effect on comprehension and also increased long-term retention of knowledge.[126] Similar studies replicate this effect on humans, and one even suggests that active engagement enhances learning in rhesus monkeys.[127]

The second type of study involves comparison of real classrooms. Because randomized, controlled studies are challenging with real students and teachers, many designs have been used. Some compare student performance in courses that are taught traditionally for many years with the same instructor using the same exams, with the only change the introduction of active exercises.[128] Others have used parallel sections of the same course,[129] and others have randomly introduced active learning into some class sessions and not others within a single course and analyzed student performance on exam questions on topics taught with active or traditional methods.[13°]

Numerous studies in chemistry, physics, biology, math, and engineering courses show that students learn, perform, and develop higher order thinking skills in active settings better than in passive ones (Table F-2). In almost any research study on humans in real world settings, concern arises that other factors co-vary with the variable of interest. Many studies control for these confounding variables as indicated below (Table F-1).

Given the size of the body of peer-reviewed research about active learning; the variation in experimental design among the studies; the diverse settings and subjects used; the consistency of findings across many STEM disciplines; and the concordance between studies of subjects under experimental conditions and studies of real STEM classes, the conclusion is convincing: teaching methods that require active engagement of the mind lead to more learning than does lecturing alone.

Table F-1. Controls for Confounding Variables in Classroom Learning Studies

Confounding Variable	Approach that Avoid this Problem
Better students in the active learning cohort	• Randomize the students in the active learning and comparison groups • Use matched groups with similar grades in previous courses
Instructor aims to prove that active learning is more effective	• Compare a traditional course taught to one by an outstanding instructoraiming to prove that active methods are *not* better than lectures alone
Active learning professor "teaches to the exam"	• Use standardized national test • Have students interviewed about the course content by a colleague who does not know which students received which treatment and who had not attended either course • Test students' ability to pose good scientific questions as judged by blind reviewers

The Evidence Summarized

Today, hundreds of papers have documented the scientific evidence regarding effective teaching,[131] including examples of large studies with robust findings in this field. A meta-analysis of 62 physics courses (14 traditional, 48 active) taught across the U.S. showed that among a total of 6,000 students, performance on a common test was higher among those who were taught with active methods.[132] Another study analyzed 39 studies of small-group learning and showed that it enhanced academic erformance, attitudes toward learning, and persistence in STEM.[133] In chemistry, a meta-analysis of controlled studies of high school chemistry, college introductory chemistry, and organic chemistry reported that nearly all of 37 studies involving 3,500 students showed statistically significant positive effects of active learning, and the average effect of active learning across all studies would move a student from the 50th percentile to the 70th percentile.[134] Among the 37 research studies reviewed, 11 also showed improvement of student attitudes toward science and 9 showed an average 22 percent higher retention of students in STEM after an active learning chemistry course than a traditional one.[135]

In the 1990s, many medical schools changed from a traditional style of delivery to problem-based learning in courses for medical students. University of Missouri-Columbia studied the impact of this change on student performance on the national Medical Licensing Examination. They found a significant improvement of scores associated with the change. For example, among the classes in 1995 and 1996, who were taught in the traditional courses, an average of 8 students per year scored in the 90th percentile. In contrast, in 1997-2000, an average of 21 students per year scored in the 90th percentile. Performance improved over time, apparently due to increased faculty experience in teaching with the problem-based style, so that by 2000, 29 students scored in the 90th percentile, representing a greater than three-fold increase compared with the traditional curriculum.[136] Subsequent studies showed that the students taught by problem-based learning methods received better evaluations from residency directors.[137]

In addition to experimental and classroom data, the enhancement of learning in active settings is supported by neurobiology and common experience. The current understanding of knowledge acquisition, short-term and long-term memory, and brain development indicate that learning changes the brain and that is accomplished by an active process of building neural connections. These are constructed through active processing.

The research indicates that many different types of active engagement can accomplish learning gains. Introduction of clickers into a lecture,[138] having students solve a problem before attending a lecture,[139] use of group discussion,[14°] problem-solving,[141] individual writing or "one-minute papers,"[142] taking a test,[143] conducting an inquiry-based lab,[144] and combinations of these activities all have had significant impacts in improving learning. Therefore, the support for using evidence-based teaching methods presented in this report is not advancing a single type of teaching. Instead, based on the numerous data points available, we posit that the key change is to bring to STEM classrooms various approaches that truly engage students intellectually and involve thinking, problem-solving, questioning, or analyzing information. Based on the weight and variety of the research evidence, it is reasonable to say that if one active-learning event in which students engaged and received feedback were incorporated into each classroom session of every introductory college class in the United States, science education would likely be transformed.

Box F-1: Enhanced Learning in a Large Physics Class

A recent experiment at the University of British Columbia demonstrated the feasibility of using active learning to greatly enhance student learning in large classes at no additional cost.

In the second term of a first-year Electricity and Magnetism course, one group of students was taught in three hours of lecture by an experienced instructor, while another group received the same material through three hours of interactive learning. Altogether, 267 students heard lectures, while 271 students were taught with a method known as "deliberate practice" based on recent findings in cognitive psychology and physics education. The instructor for the experimental group began by giving students a multiple-choice question on a particular concept. The students discussed the question in small groups and answered electronically, revealing their understanding or lack of understanding of a topic. The instructor took this feedback into account during a discussion of the topic before repeating the process with the next concept.

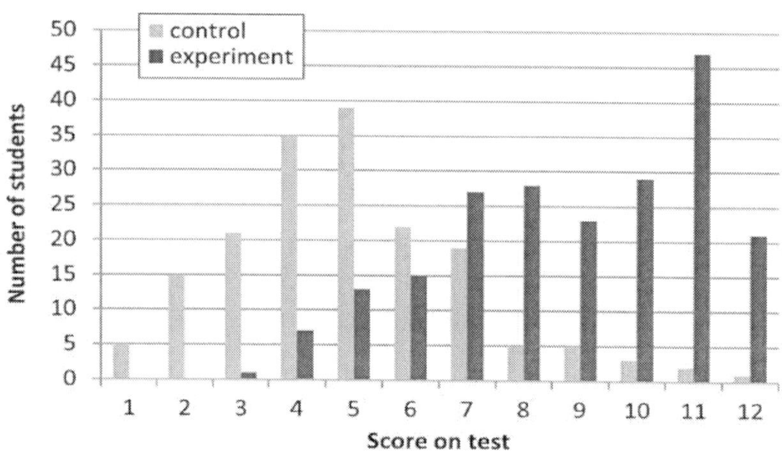

Source: Deslauriers, L., E. Schelew, and C. Wieman. (2011). "Improved learning in a large-enrollment physics class." *Science* 332: 862-864."

The goal was for students to spend as little time as possible passively listening and as much time as possible making and testing predictions and arguments, solving problems, and critiquing their reasoning and that of others.

In the non-traditional class, attendance grew from 57 to 75 percent, engagement rose from 45 to 85 percent, and the students learned twice as much based on test results as the students in the traditional section (see figure). In the traditional section, attendance and engagement remained unchanged.

In a survey afterwards, 90 percent of students in the experimental group agreed that they enjoyed the interactive teaching technique. The technique did not require additional staff or small or specialized classrooms.

APPENDIX G: REVIEW OF EVIDENCE THAT RESEARCH EXPERIENCES HAVE IMPACTS ON RETENTION

One way to engage and, therefore, retain students in STEM subjects is to involve them in contemporary, authentic research during the first two years of college (see Box G-1). For example, in a randomized trial at the University of Michigan, students who engaged in research with a professor as sophomores were much less likely to leave science majors than those who did not. Though the numbers of students involved were relatively small, the results were dramatic for all ethnic groups: attrition rates dropping from 20% to 11% for black students, from 14% to 0% for Hispanic students, and from 5.5% to 1.4% for white students.[145] A nationwide assessment of 4,500 students involved in undergraduate research found that the research experience clarified students' interests and increased their confidence.[146] Close to 70% of those surveyed said that their interest in a STEM career increased due to their experience, and about 30% of the students who had never considered earning a PhD now expected to do so. The surveys did not detect significant differences between students based on gender or demographic group. The conclusion of the researchers was that "the inculcation of enthusiasm is the key element—and the earlier the better." Additionally, an intervention of early research experience at UC-Davis showed improved grades across STEM courses and improved retention in STEM majors for students who are given rigorous academic program during their first two years of college, are funded to work in research laboratories during their sophomore year, and are provided personal support and guidance (see Box G-2).

Box G-1: The Freshman Research Initiative at Ut Austin

The Freshman Research Initiative at the University of Texas, Austin, enrolls 25 percent of the freshman class in the College of Natural Sciences in threesemesterlong laboratory courses based on faculty research programs.

The program offers firstyear students the opportunity to do cuttingedge, original, publishable research in chemistry, biochemistry, nanotechnology, molecular biology, physics, astronomy, or computer science.

The faculty member leading the course provides the overall direction for the research. Postdoctoral "Research Educators" (REs) organize the entering students' laboratory work and curricula.

Mentoring includes help with presentations, data collection and analysis, and placement after the threesemester "research stream."

Early results suggest that student retention in STEM programs is 30 to 35 percent higher for students in the initiative.

The program also has formal ways to help students continue in research in faculty laboratories, research abroad, or industry internships.

A key feature of the Freshman Research Initiative is the autonomy of the REs to work directly with students and shape their experience and motivation.

The REs are stakeholders in the institute's success and make it a hotbed for innovation in teaching.

Source: University of Texas at Austin website: http://fri.cns.utexas.edu/aboutfri.

Box G-2: Participation in Research Improves Stem Persistence and Performance

The UCDavis Biology Undergraduate Scholars Program (BUSP) Program is an intensive enrichment program for undergraduate students who have a strong interest in life science fields. BUSP, sponsored by the College of Biological Sciences at UCDavis, enriches the undergraduate experience by providing exciting and challenging opportunities to learn about and participate in the biological sciences.

BUSP students enroll in a specially designed, rigorous academic program during their first two years of college, are funded to work in a biology research laboratory during their sophomore year, and meet regularly with skilled advisers who offer academic guidance and personal support.

The Table below summarizes BUSP students' persistence and performance in STEM foundation courses, such as chemistry and calculus, for students of the underrepresented minority (URM) who participated in the BUSP program (URMBUST) as compared to students of the underrepresented minority, generally (URM comparison), or white and Asian students.

Persistence and performance in foundation courses by group.

Group	Total N	Calculus Persistence	Calculus GPA	Chemistry Persistence	Chemistry GPA
URM comparison	1,267	42%	2.69	43%	2.56
URM -BUSP	336	75%	2.94	81%	3.1
White/ Asian	5,559	54%	2.89	55%	2.79

Source: Villarejo, M., A. Barlow, D. Kogan, B.D. Veazey, and J. Sweeney (2008). "Encouraging Minority Undergraduates to Choose Science Careers: Career Paths Survey Results." CBE-Life Sciences Education 7(4): 1-16.

Appendix H: Effective Programs to Improve STEM Undergraduate Education

Building STEM Communities

Many programs have proven effective at addressing issues of retention and completion in STEM majors by focusing on building a community of STEM scholars, including the Meyerhoff Scholars Program at the University of Maryland, Baltimore County (see Box H-1), the Science Posse program that is beginning in several universities (see Box H-2), and the Louisiana Science, Technology, Engineering & Mathematics (LA-STEM) Research Scholars Program.[147] Common to all these programs is a mentoring community in which upper-division students work with beginning students to provide guidance and model success; access to research groups early in the undergraduate experience; bridge programs to prepare students for the intellectual content of the first year; and group recognition of the need to succeed in introductory and gateway courses. All of these and similar programs require funding, both for students, many of whom are receiving financial aid, and for the staff members and time needed to create and guide learning communities.

Box H-1: The Meyerhoff Scholars Program

The Meyerhoff Scholars Program at the University of Maryland, Baltimore County, has been at the forefront of efforts to increase diversity among future leaders in science, engineering, and related fields. Started in 1988, the program now has more than 1,000 alumni. Key components of the program include scholarships contingent on maintaining a B-average in STEM majors, an intensive six-week summer bridge program, a family-like program community, an emphasis on achieving at the highest levels, personal advising and counseling from program staff, summer research internships in national and international laboratories, science mentoring, and support from administrators and faculty.

The nomination-based application process is open to prospective undergraduate students of all backgrounds who plan to pursue doctoral study in the sciences or engineering and who are interested in the advancement of minorities in those fields. The program's success is built on the premise that, among like-minded students who work closely together, positive energy is contagious. By assembling such a high concentration of high-achieving students in a tightly knit learning community, students continually inspire one another to do better.

Among African American students who entered the program between 1996 and 2003, 51% (88 of 172) attended STEM PhD and MD/PhD programs. An additional 40% entered master's programs, particularly in technical fields, or medical school.

Many representatives from Federal agencies, campuses, and corporations across the country have visited UMBC's campus to learn more about the program's success. The College Board's National Task Force on Minority High Achievement has praised the Meyerhoff Scholars Program as a model that provides lessons that could be broadly applied.

Source: University of Maryland, Baltimore County website: http://umbc.edu /meyerhoff/.

Engaging and Preparing Rising College Students

Bridge programs, which are typically offered in the summer between high school graduation and the first term of college, can help prepare entering students for the rigors of college academics and life (see Boxes H-2 and H-3). Typically, high school juniors and seniors live on campus and receive classroom instruction, research experience, career counseling, SAT and ACT preparation, and mentoring from graduate students and faculty. Most of these programs, such as Carnegie Mellon University's Summer Academy for Mathematics and Science and the California State Summer School for Mathematics and Science, are open to high school students on statewide or nationwide basis. Some are aimed at the underrepresented majority to provide incoming students with the intellectual, personal, and social supports they will need to excel.

Box H-2: A Posse Program for Stem Fields

The Posse Foundation is a successful college access and youth leadership development program. Through creative partnerships between local communities and 39 select colleges and universities, Posse currently recruits, nurtures, and delivers outstanding student leaders from eight urban sites: Atlanta, Boston, Washington D.C., Chicago, Los Angeles, Miami, New Orleans, and New York.

Since its inception, Posse has sent 3,650 students to college, 90 percent of whom have graduated. As one measure of the program's impact on leadership development, over 70 percent of Posse Scholars either start new campus organizations or become presidents of existing ones. A recent survey shows that more than 45 percent of Posse alumni either have completed a graduate degree or are currently in graduate school.

Posse's college access process is noteworthy. Each fall seniors are nominated by high schools and community-based organizations in the eight cities. Posse staff and volunteers evaluate students, looking for leaders with true commitment and potential. Partnering colleges and universities then select ten-student Posses in December of the students' senior year of high school. During the remainder of their senior year, the students participate in weekly sessions with staff trainers and peers who provide scholastic and cultural preparation for college. Once on campus the students are mentored by staff and upperclassmen. The home community supports the recruiting process, and the partner colleges and universities provide four-year scholarships. Posse recently began a STEM Posse initiative on three campuses based on the proven elements of the original program, with additional components needed for STEM.

Thus far Brandeis University, the University of Wisconsin at Madison, and Franklin and Marshall College are admitting STEM Posses. The program identifies students with an interest in STEM and provides extra pre-collegiate training during their senior year, a two-week campus immersion program just prior to matriculation, intensive mentoring in STEM-related areas, and placement in research opportunities throughout the four undergraduate years.

Although highly successful, growth of the program is limited by the financial burden on the participating institutions. The reach of this program could be greatly enhanced by a Federal or other partner cost-sharing program with the schools.

Since 75 to 80 percent of the students require financial aid, a 50 percent Federal or other partner contribution to the scholarships would clearly allow more institutions to participate by relieving their financial aid budgets and would target federal financial aid dollars to a group of students with high potential for success.

Source: Posse Foundation website: http://www.possefoundation.org/.

Box H-3: Mit Helps Minority High School Students Succeed in College Stem Majors

A three-week engineering program for minority high school students at MIT that began in 1974 has evolved into a national model for widening the pipeline of underrepresented college graduates in STEM fields.

Today the Minority Introduction To Engineering and Science (MITES) program supports 60 to 80 high school students, annually, in the summer after their junior year.

They live in an MIT dormitory for a six-andone-half week program of academic work, confidence-building, and development of learning-to-learn skills.

Of the 1,765 alumni of the program to date, 34 percent (more than 600 students) have gone on to MIT. Recent MITES alumni have also gone to Harvard University, Stanford University, and other exceptional schools. The MITES alumni have been found to be consistently strong performers in college. At MIT, the graduation rates of MITES alumni are 12 percentage points higher than the graduation rates of minority students who did not attend MITES. MITES students at MIT also graduate with grade point averages comparable to the majority MIT student population. Because of its reputation and systematic outreach, the program receives some 500 to 700 applications from around the country, making it more selective than MIT itself. In the summer of 2010, 65 students were selected from 22 states and Puerto Rico. Acceptance includes consideration of a student's status as first generation college and those who lack a family members background in science and engineering. Upon arrival, students are tested to establish individual benchmarks and to guide course selection. Through evaluation updates, instructors write detailed evaluation of each student's mastery of the subject in relation to his or her benchmark. Students are given many assignments and quizzes but no final exam or final grades. The curriculum uses the cultural context—having students from different minority groups living and working together—to show how cultural diversity and academic achievement can be connected. The MITES program is entirely scholarship-based. Support comes from dozens of companies and foundations, including 3M Worldwide, Boeing, and the Broad Institute of MIT and Harvard, and from alumni.

Source: Massachusetts Institute of Technology website: http://web.

Partnerships between Two-Year Colleges and Four-Year Colleges

Collaborative partnerships between two-year colleges and four-year institutions would provide greater access to and opportunities for advanced STEM education to a growing number of students (see Box H-4).

Box H-4: Articulation between Two-Year and Four-Year Institutions

A keystone of the applied STEM manufacturing skills certification model at the Lorain County Community College (LCCC) in Cleveland, Ohio, is a unique partnership with four-year institutions. LCCC is the only community college in the state that offers a program enabling individuals to earn Bachelor's and Master's degrees from any of eight Ohio universities without leaving the LCCC campus.

The University Partnership program facilitates seamless, STEM-related education and career pathways for students completing manufacturing-based programs at the Associate's- and applied science- level. Programs articulate with a variety of Bachelor's of Science degrees in engineering and engineering technology for students who want to pursue additional levels of higher education. As part of the industry certification initiative, college leaders launched a review of the curriculum's alignment with industry requirements. Faculty identified new or revised content to address skill requirements. The Manufacturing Advocacy and Growth Network (MAGNET), an employer-led organization, held employer meetings to validate the certification pathways and discuss embedded skills, including both applied STEM and critical "soft" skills. The University Partnership at LCCC enables students to gain the depth and breadth of applied STEM skills required to spur innovation and creativity in the modern workplace.

Source: Lorain County Community College website: http://www.lorainccc. edu/up.

Partnerships between Minority-Serving Institutions and Other Colleges and Universities

Minority-serving institutions (MSIs) can serve as key intermediaries to improve the numbers, preparation, and diversity of students interested in STEM fields.'[48] Collaborative efforts between MSIs and other colleges and universities could greatly improve educational experiences in STEM disciplines (see Box H-5).

Box H-5: A Successful Partnership between a Historically Black Teaching-Focused College and a Research University

Institutional collaborations that benefit both partners are exemplified by the joint endeavor developed by the University of New Hampshire (UNH) and Elizabeth City State University (ECSU), which are a research university and a teaching-focused historically black institution, respectively. The goal of the partnership was to expand the interest and success of students from underrepresented groups entering STEM careers through expanded scientific knowledge and enhanced educational opportunities. The collaboration involved exchanges of students and faculty, development of new courses, co-teaching, and joint faculty meetings and presentations. Specific outcomes were providing UNH students with a more diverse educational environment, ECSU students with access to research labs, and both campuses with Federal support for improved STEM research and education.

The collaboration has delineated a set of best practices that could be useful to other alliances, including:

- Institutional commitment and faculty engagement
- Mutual respect and shared time commitments
- An engaged leader
- Critical change agents
- Initiation of difficult dialogues
- Preparing for growth and evolution

Source: Williams, J.E., C. Wake, E. Abrams, G. Hurtt, B. Rock, K. Graham, S. Hale, L. Hayden, W. Porter, R. Blackmon, M. LeCompte, and D. Johnson.(2011). "Building a model of collaboration between historically black and historically white universities." *Journal of Higher Education Outreach and Engagement* 15(2): 35-56.

Partnerships between Higher Education and Business

Some U.S. businesses have found effective ways to partner to enhance STEM education and career-readiness in high schools, colleges, and universities (see Boxes H-6 and H-7). Involvement of the private sector in training of the future workforce can provide motivation and confidence for students in their ability to perform a STEM-capable job, enhanced training and useful experience, and career readiness.

BOX H-6: EMT Summer Academy

Foothill College in Los Altos Hills, CA, offers an accelerated Emergency Medical Technician (EMT) Summer Academy in partnership with the Silicon Valley Community Collaborative (SVCC), the Central County Occupational Center (CCOC), and the San Jose Job Corps. The EMT Academy is presented as a stepping stone for students' advancement in allied health and medical careers.

In addition to meeting labor force needs, this program is designed to serve as a model for increasing the retention of underrepresented students in community colleges, particularly in STEM-related fields.

The central components of the program include EMT certification, career and college counseling, tutoring, supported transition to EMT employment and/or college programs, removal of barriers in navigating institutional bureaucracy, and implementation of engagement strategies.

Source: Foothill College website: http://www.foothill.edu/bio/programs/emt/.

Box H-7: Harrisburg University for Science and Technology

In Harrisburg, Pennsylvania, a postsecondary institution is helping students who leave high school without good preparation become marketable in STEM fields.

The Harrisburg University for Science and Technology (HU), which has grown from 100 to 722 students since 2005, is a private university with the mission of readying the central Pennsylvania workforce for 21st century jobs.

Just 12 percent of residents in the Harrisburg area have a college degree, and area colleges are under producing STEM degrees as compared with similar regions. As manufacturing companies have closed, the local economy needs more skilled STEM workers to revive.

The HU academic format is interdisciplinary, without departments or tenure. Courses are organized around learning objectives, and corporate partners advise on course design. Communication and teamwork are stressed throughout the curriculum.

Two thirds of the students are adults, many sponsored by their employers. All students are coached on life issues such as time management and juggling family and careers. An executive search firm helps new students define career paths, and each has a business mentor. Each student builds an "e-portfolio" that includes performance, comments from faculty, and measures of civic engagement.

Of its original 100 students, 92 were hired into the fields they studied, with salaries of $50,000 to $60,000 per year, according to Mel Schiavelli, President of HU. Another striking result is that employers of 18-22 year old students say they do not have to spend 12 to 18 months teaching their new hires how to fit into corporate culture. The students were already mentored through internships and academic-year projects based on workplace needs.

Besides helping students and employers HU is helping to revive downtown Harrisburg, with a new building and dormitory and $30 million in annual economic impact.

Source: Based on PCAST Working Group on Undergraduate STEM education discussions with Mel Shiavelli, President, Harrisburg University for Science and Technology, May 2011, and data from Harrisburg University of Science and Technology website.

Appendix I: References for Tables 2, 3, and 4

References for Table 2

Almer, E., K. Jones, and C. Moeckel. (1998). "The impact of one-minute papers on learning in an introductory accounting course." *Issues in Accounting Education* 13(3): 485-495.

Anderson, W. L., Mitchell, S. M., and M.P. Osgood. (2005). "Comparison of student performance in cooperative learning and traditional lecture-based biochemistry classes." *Biochemistry and molecular biology education: A bimonthly publication of the International Union of Biochemistry and Molecular Biology* 33(6): 387-93.

Armbruster, P., M. Patel, E. Johnson, and M. Weiss. (2009). "Active learning and student-centered pedagogy improve student attitudes and performance in introductory biology." *Education* 8: 203-213.

Armstrong, N., S. Chang, and M. Brickman. (2007). "Cooperative learning in industrial-sized biology classes." *Education* 6: 163-171.

Beichner, R. J., J.M. Saul, D.S. Abbott, J.J. Morse, D.L. Deardorff, R. J. Allain, et al. (2007). "The student-centered activities for large enrollment undergraduate programs (SCALE-UP) project." In E. F. Redish & P. J. Cooney (Eds.), *Research-Based Reform of University Physics*. College Park, MD: American Association of Physics Teachers.

Born, W. K., W. Revelle, and L.H. Pinto. (2002). "Improving biology performance with workshop groups." *Science Education* 11(4): 347.

Buckley, M., H. Kershner, K. Schindler, C. Alphonce, and J. Braswell. (2004). "Benefits of using socially-relevant projects in computer science and engineering education." *SIGCSE '04: Proceedings of the 35th SIGCSE Technical Symposium on Computer Science Education* 36(1): 482–486.

Capon, N. and D. Kuhn. (2004). "What‘s so good about problem-based learning?" *Cognition and Instruction* 22(1): 61-79.

Chizmar, J. F. and A.L. Ostrosky. (1998). "The one-minute paper: Some empirical findings." *The Journal of Economic Education* 29(1):3.

Cortright, R. N., H.L. Collins, D.W. Rodenbaugh, and S.E. DiCarlo. (2003). "Student retention of course content Is improved by collaborative-group testing." *AJP: Advances in Physiology Education* 27(3): 102-108.

Cortright, R. N., H.L. Collins, and S.E. DiCarlo. (2005). "Peer instruction enhanced meaningful learning: ability to solve novel problems." *Advances in Physiology Education* 29(2): 107-11.

Crouch, C. H., and Mazur, E. (2001). "Peer Instruction: Ten years of experience and results." *American Journal of Physics* 69(9): 970.

Fagen, A. P. (2002). "Peer instruction: results from a range of classrooms." *The Physics Teacher* 40(4): 206.

Fonseca, A. P., C.I. Extremina, and A.F. Fonseca. (2004). "Concept mapping: A strategy for meaningful learning in medical microbiology." *First International Conference on Concept Mapping*. Pamplona, Spain.

Freeman, S., E. O. Connor, J.W. Parks, M. Cunningham, D. Hurley,D. Haak, C. Dirks, and M.P. Wenderoth. (2007). "Prescribed active learning increases performance in introductory biology." *Education* 6:132-139.

Harris, M. A., R.F. Peck, S. Colton, J. Morris, E.C. Neto, and J. Kallio. (2009). "A combination of hand-held models and computer imaging programs helps students answer oral questions about molecular structure and function: A controlled investigation of student learning." *CBE—Life Sciences Education* 8(1): 29-43.

Klappa, P. (2009). "Promoting active learning through "pub quizzes"— a case study at the University of Kent." *Evaluation* 14 (December), Article C2.

Lasry, N., E. Mazur, and J. Watkins. (2008). "Peer instruction: From Harvard to the two-year college." *American Journal of Physics* 76(11): 1066.

Lewis, S. E. and J.E. Lewis. (2005). "Departing from lectures: An evaluation of a peer-led guided inquiry alternative." *Journal of Chemical Education* 82(1): 135.

McDaniel, C. N., B.C. Lister, M.H. Hanna, and H. Roy. (2007). "Increased learning observed in redesigned introductory biology course that employed web-enhanced, interactive pedagogy." *CBE—Life Sciences Education* 6: 243-249.

McDaniel, M., H. Roediger, and K. McDermott. (2007). "Generalizing test-enhanced learning from the laboratory to the classroom." *Psychonomic Bulletin & Review* 14(2): 200-206.

O Sullivan, D. W. and C.L. Cooper. (2003). "Evaluating active learning A new initiative for a general chemistry curriculum." *Journal of College Science Teaching* 32(7): 448-453.

Pelaez, N. J. (2002). "Problem-based writing with peer review improves academic performance in physiology." *Advances in Physiology Education* 26(3): 174-184.

Preszler, R. (2004). "Cooperative concept mapping: Improving performance in undergraduate biology." *Journal of College Science Teaching* 33(6): 30-35.

Preszler, R. W., A. Dawe, and C.B. Shuster. (2007). "Assessment of the Effects of Student Response Systems on Student Learning and Attitudes over a Broad Range of Biology Courses." *CBE—Life Sciences Education* 6(1): 29 - 41.

Preszler, R. W. (2009). "Replacing lecture with peer-led workshops improves student learning." *CBE—Life Sciences Education* 8: 182-192.

Rivard, L.P. and S.B. Straw. (2000). "The effect of talk and writing on learning science: An exploratory study." *Science Education* 84: 566-593.

Schwartz, D. L. and J.D. Bransford (1998). "A time for telling." *Cognition & Instruction* 16: 475-522.

Smith, M., W. Wood, W. Adams, C. Wieman, J. Knight, N. Guild, et al. (2009). "Why peer discussion improves student performance on in-class concept questions." *Science* 323: 122-124.

Smith, M.K., W.B. Wood, K. Krauter, and J.K. Knight. (2011) "Combining Peer Discussion with Instructor Explanation Increases Student Learning from In-Class Concept Questions." *CBE—Life Sciences Education* 10: 55-63.

Steele, J. E. (2003). "Effect of essay-style lecture quizzes on student performance on anatomy and physiology exams." *Bioscene: Journal of College Biology Teaching* 29(4): 15-20.

Tessier, J. (2004). "Using peer teaching to promote learning in biology." *Journal of College Science Teaching* 33(6): 16-19.

Tessier, J. (2007). "Small-group peer teaching in an introductory biology classroom." *Journal of College Science Teaching* 36(4): 64-69.

Tien, L.T., V. Roth, and J.A. Kampmeier. (2002). "Implementation of a peer-led team learning instructional approach in an undergraduate organic chemistry course." *Journal of Research in Science Teaching* 39(7): 606-632.

Traver, H. A., M.J. Kalsher, J.J. Diwan, and J. Warden. (2001). "Student reactions and learning: Evaluation of a biochemistry course that uses web technology and student collaboration." *Biochemistry and Molecular Biology Education* 29: 50-53.

Yarden, H., G. Marbach-ad, and J.M. Gershoni. (2004). "Using the concept map technique in teaching introductory cell biology to college freshmen." *Bioscene: Journal of College Biology Teaching* 30(1): 3-13.

References for Table 3

American Association for the Advancement of Science. (2011). "Vision and Change in Undergraduate Science Eduction: A View for the 21st Century." Washington, DC.

Austin, A. E., M.R. Connolly, and C.L. Colbeck (2008). "Strategies for Preparing Integrated Faculty: The Center for the Integration of Research, Teaching, and Learning." *New Directions for Teaching and Learning* 113: 69-81.

Barlow, A. and M. R. Villarejo (2004). "Making a difference for minorities: Evaluation of an educational enrichment program." *Journal of Research in Science Teaching* 41(9): 861-881.

Bartlett, K. (2003). "Towards a true community of scholars: Undergraduate research in the modern university." *Journal of Molecular Structure: THEOCHEM* 666-667: 707-711.

Beatty, I. D. (2004). "Transforming student learning with classroom communication systems." *Educause Center for Applied Research: Research Bulletin* 2004(3): 1-13.

Bouwma-Gearhart, J. L., S.B. Millar, S.S. Barger, and M.R. Connolly (2007). "Doctoral and Postdoctoral STEM Teaching-related Professional Development: Effects on the Early Career." *American Educational Research Association Annual Meeting.*

Burstyn, J., S. Sellers, A. Cabrera, K. Freidrich, and L. Giovanetto. (2006). "Resources for Inclusive Teaching in Science, Technology, Engineering and Mathematics." CIRTL Diversity Institute Literature Review. Accessible at http://www.cirtl.net/bibliography.

Caldwell, J. E. (2007). "Clickers in the Large Classroom: Current Research and Best-Practice Tips." *CBE— Life Sciences Education* 6(1): 9-20.

Campbell, S. P., A. K. Fuller, and D.A.G. Patrick. (2005). "Looking beyond research in doctoral education." *Frontiers in Ecology and the Environment* 3(3): 153-160.

Carter, F. D., M. Mandell, and K.I. Maton. (2009). "The Influence of On-Campus, Academic Year Undergraduate Research on STEM Ph.D. Outcomes: Evidence From the Meyerhoff Scholarship Program." *Educational Evaluation and Policy Analysis* 31(4): 441-462.

Connolly, M. R. (2008). "Effects of a Future-Faculty Professional Development Program on Doctoral Students and Postdocs in Science, Technology, Engineering and Math: Findings from a Three-year Longitudinal Study." *Presented at conference on Preparing for Academic Practice: Disciplinary Perspectives*. Oxford, England.

Donofrio, L. A., B. Russell, et al. (2007). "Mentoring. Linking student interests to science curricula." *Science* 318(5858): 1872-1873.

Felder, R. M., A. Rugarcia, and J.E. Stice. (2000). "The future of engineering education: Assessing teaching effectiveness and educational scholarship." *Chemical Engineering Education* 34(3): 198–207.

Gainen, J. (1995). "Barriers to success in quantitative gatekeeper courses." *New Directions for Teaching and Learning* 61: 5–14.

Gilmer, T. C. (2007). "An Understanding of the Improved Grades, Retention and Graduation Rates of STEM Majors at the Academic Investment in Math and Science (AIMS) Program of Bowling Green State University (BGSU)." *Higher Education* 8: 11-21.

Haak, D. C., J. HilleRisLambers, E. Pitre, and S. Freeman. (2011). "Increased Structure and Active Learning Reduce the Achievement Gap in Introductory Biology." *Science* 332: 1213-1216.

Hathaway, R., B.A. Nagda, and S. Gregerman. (2002). "The relationship of undergraduate research participation to graduate and professional education pursuit: An empirical study." *Journal of College Student Development* 43(5): 614-631.

Haudek, K. C., J. J. Kaplan, et al. (2011). "Harnessing Technology to Improve Formative Assessment of Student Conceptions in STEM: Forging a National Network." *CBE—Life Sciences Education* 10(2): 149-155.

Hunter, A-B., S. L. Laursen, and E. Seymour. (2007). "Becoming a scientist: The role of undergraduate research in students'cognitive, personal, and professional development." *Science Education* 91: 36-74.

Kight, S.L., J.J. Gaynor, and S.D. Adams. (2006). "Undergraduate Research Communities: A Powerful Approach to Research Training." *Journal of College Science Teaching* July/August, 34-39.

Kinkel, D. H. and S. E. Henke. (2006). "Impact of undergraduate research on academic performance, educational planning, and career development." *Journal of Natural Resources and Life Sciences Education* 35: 194-201.

Lockwood, P. (2006). "Someone like me can be successful: Do college students need same-gender role models?" *Psychology of Women Quarterly* 30(1): 36-46.

Lopatto, D., C. Alvarez, et al. (2008). "Undergraduate research. Genomics Education Partnership." *Science* 322(5902): 684-685.

May, G. S. and D. E. Chubin (2003). "A retrospective on undergraduate engineering success for underrepresented minority students." *Journal of Engineering Education* 92(1): 33.

Miller, S., C. Pfund, C. Maidle Pribbenow, and J. Handeslman. (2008). "Scientific Teaching in Practice." *Science* 322: 1329-1330.

Muller, C. B. (1997). "The potential of industrial 'E-Mentoring'as a retention strategy for women in science and engineering." *Frontiers in Education Conference1997 - 27th Annual Conference, Teaching and Learning in an Era of Change.* Proceedings *2:* 622-626.

Ong, A. D., J. S. Phinney, et al. (2006). "Competence under challenge: Exploring the protective influence of parental support and ethnic identity in Latino college students." *Journal of Adolescence* 29(6): 961-979.

Ovando, M. N. (1994). "Constructive Feedback: A Key to Successful Teaching and Learning." *International Journal of Educational Management* 8(6): 19 - 22.

Peckham, J., P. Stephenson, J-Y Hervé, R. Hutt, and Miguel Encarnação. (2007). "Increasing student retention in computer science through research programs for undergraduates." *SIGCSE '07: Proceedings of the 38th SIGCSE Technical Symposium on Computer Science Education* 39(1): 124–128.

Pfund, C., S. Miller, K. Brenner, P. Bruns, A. Chang, D. Ebert-May, A.P Fagen, J. Gentile, S. Gossens, I.M. Khan, J.B. Labov, C.M. Pribbenow, M. Susman, L. Tong, R. Wright, R.T. Yuan, W.B. Wood, and J. Handelsman, J. (2009). "Summer Institute to Improve University Science Teaching." *Science* 324: 470-471.

Rodriguez, A. L., F. Guido-DiBrito, V. Torres, and D. Talbot. (2000). "Latina College Students: Issues and Challenges for the 21st Century." *NASPA Journal* 37(Spring 2000).

Russell, S. H., M.P. Hancock, and J. McCullough (2007). "The pipeline. Benefits of undergraduate research experiences." *Science* 316(5824): 548-9.

Shaffer, C. D., C. Alvarez, et al. (2010). "The Genomics Education Partnership: Successful Integration of Research into Laboratory Classes at a Diverse Group of Undergraduate Institutions." *CBE—Life Sciences Education* 9(1): 55-69.

Springer, L., M. E. Stanne, and S.S. Donovan. (1999). "Effects of small-group learning on undergraduates in science, mathematics, engineering, and technology: A meta-analysis." *Review of Educational Research* 69(1): 21-51.

Stout, J. G. D., Nilanjana; Hunsinger, Matthew; McManus, Melissa A. (2011). "STEMing the tide: Using ingroup experts to inoculate women's self-concept in science, technology, engineering, and mathematics (STEM)." *Journal of Personality and Social Psychology* 100: 255-270.

Summers, M. F. and F. A. Hrabowski (2006). "Preparing Minority Scientists and Engineers." *Science* 311(5769): 1870-1871.

Sy, S. R. a. R., J. (2008). "Family Responsibilities Among Latina College Students From Immigrant Families." *Journal of Hispanic Higher Education* 7(3): 212-227.

Turner, P., L. Petzold, A. Shiflet, I. Vakalis, K. Jordan, and S. St. John. (2011). "Undergraduate Computational Science and Engineering Education." Society for Industrial and Applied Mathematics Review 53(3): 561–574.

University of Texas, Austin. (2011). Course Transformation Site from http://www.utexas.edu/academic/ctp/resources/literature-review /#instructionaldevtech.

Walton, G. M. and G. L. Cohen. (2011). "A Brief Social-Belonging Intervention Improves Academic and Health Outcomes of Minority Students." *Science* 331(6023): 1447-1451.

Wei, C. A. and T. Woodin. (2011). "Undergraduate Research Experiences in Biology: Alternatives to the Apprenticeship Model." *CBE—Life Sciences Education* 10(2): 123-131.

Wild, L. and L. Ebbers. (2002). "Rethinking student retention in community colleges." *Community College Journal of Research and Practice* 26: 503- 519.

Yoon, K. S., T. Duncan, et al. (2007). "Reviewing the evidence on how teacher professional development affects student achievement." *Issues & Answers Report, REL, Washington, DC: U.S. Department of Education, Institute of Education Sciences, National Center for Education Evaluation and Regional Assistance, Regional Educational Laboratory Southwest.* 2007 (033).

References for Table 4

Barlow, A. and M. R. Villarejo (2004). "Making a difference for minorities: Evaluation of an educational enrichment program." *Journal of Research in Science Teaching* 00: 1-2.

Carter, F. D., M. Mandell, and K.I. Maton. (2009). "The Influence of On-Campus, Academic Year Undergraduate Research on STEM Ph.D. Outcomes: Evidence From the Meyerhoff Scholarship Program." *Educational Evaluation and Policy Analysis* 31(4): 441-462.

Foertsch, J. A., B. B. Alexander, and D.L. Penberthy. (1997). "Evaluation of UW-Madison's summer undergraduate research programs (final report)." Madison, WI: University of Wisconsin-Madison, LEAD Center.

Gilmer, T. C. (2007). "An Understanding of the Improved Grades, Retention and Graduation Rates of STEM Majors at the Academic Investment in Math and Science (AIMS) Program of Bowling Green State University (BGSU)." *Higher Education* 8: 11-21.

Hathaway, R., B.A. Nagda, and S. Gregerman. (2002). "The relationship of undergraduate research participation to graduate and professional education pursuit: An empirical study." *Journal of College Student Development* 43(5): 614-631.

Hunter, A-B., S. L. Laursen, and E. Seymour. (2007). "Becoming a scientist: The role of undergraduate research in students'cognitive, personal, and professional development." *Science Education* 91: 36-74.

Junge, B., C. Quiñones, J. Kakietek, D. Teodorescu, and P. Marsteller. (2010). "Promoting Undergraduate Interest, Preparedness, and Professional Pursuit in the Sciences: An

Outcomes Evaluation of the SURE Program at Emory University." *CBE—Life Sciences Education* 9(2): 119-132.

Kight, S.L., J.J. Gaynor, and S.D. Adams. (2006). "Undergraduate Research Communities: A Powerful Approach to Research Training." *Journal of College Science Teaching* July/August, 34-39.

Kinkel, D. H. and S. E. Henke. (2006). "Impact of undergraduate research on academic performance, educational planning, and career development." *Journal of Natural Resources and Life Sciences Education* 35: 194-201.

Lopatto, D. (2004). "Survey of Undergraduate Research Experiences (SURE): first findings." *Cell Biology Education* 3(4): 270-277.

Lopatto, D. (2007). "Undergraduate research experiences support science career decisions and active learning." *CBE—Life Sciences Education* 6: 297-306.

Russell, S. H., M.P. Hancock, and J. McCullough. (2007). "The pipeline. Benefits of undergraduate research experiences." *Science* 316(5824): 548-9.

Summers, M. F. and F. A. Hrabowski. (2006). "Preparing Minority Scientists and Engineers." *Science* 311(5769): 1870-1871.

End Notes

[1] See Appendix C.

[2] See Appendix D.

[3] Lacey, T. A. and B. Wright. (2009). "Occupational employment projections to 2018." *Monthly Labor Review* 132(11):82-123.

[4] See Appendix D.

[5] Langdon, D., G. McKittrick, D. Beede, B. Khan, and M. Doms. (2011). "STEM: Good Jobs Now and for the Future." *ESA Issue Brief* #03-11. Washington, DC: U.S. Department of Commerce.

[6] Carnevale, A.P., N.Smith, and J. Strohl. (2010). *Help Wanted: Projections of Jobs and Education Requirements through 2018*. Washington, DC: Georgetown University Center on Education and the Workforce.

[7] Carnevale, A.P, N. Smith, and M. Melton. (2011). *STEM*. Washington, DC: Georgetown University Center on Education and the Workforce.

[8] U.S. General Accounting Office. (2005). *Higher Education: Federal Science, Technology, Engineering, and Mathematics Programs and Related Trends*. Washington, DC.

[9] Langdon, et al. (2011), op. cit.

[10] Scott, J. and A. Balakrishnan. "STEM Workforce Trends, Projections, and Skills Assessments." Unpublished analysis for PCAST Working Group on Postsecondary STEM Education, May 9, 2011. Washington, DC: IDA Science and Technology Policy Institute.

[11] Carnevale, A. P., and S. J. Rose. (2011). *The Undereducated American*. Washington, DC: Georgetown University Center on Education and the Workforce.

[12] Goldin, C., and L.F. Katz. (2008). *The Race Between Education and Technology*. Cambridge, MA: Harvard University Press.

[13] Radford, A.W., L. Berkner, S.D. Wheeless, and B. Shepherd. (2010). "Persistence and Attainment of 2003– 04 Beginning Postsecondary Students: After 6 Years (NCES 2011-151)." U.S. Department of Education. Washington, DC: National Center for Education Statistics. Accessible at http://nces.ed.gov/pubsearch.

[14] See Appendix C.

[15] Seymour, E. and N. M. Hewitt. (1997). *Talking about leaving*. Boulder, CO: Westview Press.

[16] Brainard, S. and L. Carlin. (1998). "A six-year longitudinal study of undergraduate women in engineering and science." *Journal of Engineering Education* 87(4):), 369-375.

[17] The concept of the "underrepresented majority" has particularly been championed by Shirley Jackson, president of Rensselaer Polytechnic Institute. For example: Jackson, Shirley. (2004). "The Perfect Storm: A Weather Forecast." Address to the annual meeting of the American Association for the Advancement of Science, Seattle, WA.

[18] Felder, R. M., G.N. Felder, and E.J. Dietz. (1998). "A longitudinal study of engineering student performance and retention v. comparisons with traditionally-taught students." *Journal of Engineering Education* 87(4): 469-480.

[19] Walton, G. M. and G. L. Cohen. (2011). "A Brief Social-Belonging Intervention Improves Academic and Health Outcomes of Minority Students." *Science* 331(6023): 1447-1451.

[20] Nagda, B. A., S. R. Gregerman, J. Jonides, W. von Hippel, and J.S. Lerner. (1998). "Undergraduate student-faculty research partnerships affect student retention." *The Review of Higher Education* 22: 55-72.

[21] Ohland, M. W., C.E. Brawner, M.M. Camacho, R.A. Layton, R.A. Long, S.M. Lord, and M.H. Wasburn. (2011). "Race, gender, and measures of success in engineering education." *Journal of Engineering Education* 100: 225-252.

[22] The National Academies (2010). *Expanding Minority Participation: America's Science and Technology Talent at the Crossroads*. Washington, DC: National Academy Press.

[23] See Appendix C.

[24] Tobias, S. (1990). *They're Not Dumb, They're Different: Stalking the Second Tier*. Tucson, AZ: Research Corporation.

[25] Seymour and Hewitt. (1997). op. cit.

[26] Tobias. (1990). op. cit.

[27] Sevo, R. (2009). "The Talent Crisis in Science and Engineering." In B. Bogue & E. Cady (Eds.). *Applying Research to Practice (ARP) Resources*. Accessible at http://www.engr.psu.edu/AWE/ARPResources.aspx.

[28] Felder, R. M., G.N. Felder, and E.J. Dietz. (1998). "A longitudinal study of engineering student performance and retention v. comparisons with traditionally-taught students." *Journal of Engineering Education* 87(4): 469-480.

[29] Nagda, B. A., S. R. Gregerman, J. Jonides, W. von Hippel, and J.S. Lerner. (1998). "Undergraduate student-faculty research partnerships affect student retention." *The Review of Higher Education* 22: 55-72.

[30] Marx, D. M. and J. S. Roman. (2002). "Female Role Models: Protecting Women's Math Test Performance." *Personality and Social Pychology Bulletin* 28: 1183-1193.

[31] Lockwood, P. (2006). "Someone like me can be successful: Do college students need same-gender role models?" *Psychology of Women Quarterly* 30(1): 36-46.

[32] Cheryan, S., J.O. Siy, M. Vichayapai, B.J. Drury and S. Kim. (2011). "Do Female and Male Role Models Who Embody STEM Stereotypes Hinder Women's Anticipated Success in STEM?" *Social Psychological and Personality Science* 2: 656-664.

[33] The National Academies. (2010). op. cit.

[34] Sy, S. R. and J. Romero. (2008). "Family Responsibilities Among Latina College Students From Immigrant Families." *Journal of Hispanic Higher Education* 7(3): 212-227.

[35] Ethier, K. A. and K. Deaux. (1994). "Negotiating Social Identity When Contexts Change—Maintaining Identification And Responding To Threat." *Journal of Personality and Social Psychology* 67: 243-251.

[36] Aronson, J. M. (Ed.) (2002). *Improving academic achievement: Impact of psychological factors in education*. San Diego, CA: Academic Press.

[37] Byars-Winston, A., Y. Estrada, C. Howard, D. Davis, J. Zalapa. (2010). "Influence of social cognitive and ethnic variables on academic goals of underrepresented students in science and engineering: A multiple-groups analysis." *Journal of Counseling Psychology* 57: 205-218.

[38] Rosenthal, H.E. and R.J. Crisp. (2006). "Reducing stereotype threat by blurring intergroup boundaries." *Personality and Social Psychology Bulletin 32*, 501–511.

[39] Kost-Smith, L. E., S.J. Pollock, N.S. Finkelstein, G.L. Cohen, T.A. Ito, and A. Miyake. (2011). "Replicating a Self-Affirmation Intervention to Address Gender Differences: Successes and Challenges." Research paper. Boulder, CO: Physics Education Research @ Colorado.

[40] Walton, G. M. and G. L. Cohen (2011). "A Brief Social-Belonging Intervention Improves Academic and Health Outcomes of Minority Students." *Science* 331(6023): 1447-1451.

[41] Lopatto, D. (2007). "Undergraduate research experiences support science career decisions and active learning." *CBE Life Sciences Education* 6: 297-306.

[42] Bartlett, K. (2003). "Towards a true community of scholars: undergraduate research in the modern university." *Journal of Molecular Structure: THEOCHEM* 666-667: 707-711.

[43] Espinosa, L. L. (2011). "Pipelines and Pathways: Women of Color in Undergraduate STEM Majors and the College Experiences That Contribute to Persistence." *Harvard Educational Review* 81(2): 209-240.

[44] Estrada, M., A. Woodcock, P.R. Hernandez, and P. Schultz, and P.W. Schultz. (2011). "Toward a model of social influence that explains minority student integration into the scientific community." *Journal of Educational Psychology* 103(1): 206-222.

[45] Anaya, G. (2001). "Correlates of performance on the MCAT: an examination of the influence of college environments and experiences on student learning." *Advances in Health Sciences Education Theory and Practice* 6: 179-191.

[46] Gregerman, S. R. (1999). "Improving the Academic Success of Diverse Students Through Undergraduate Research." *Council on Undergraduate Research Quarterly.*

[47] Hathaway, R., B.A. Nagda, and S. Gregerman. (2002). "The relationship of undergraduate research participation to graduate and professional education pursuit: An empirical study." *Journal of College Student Development* 43(5): 614- 631.

[48] Bartlett. (2003). op. cit.

[49] Kight, S.L., J.J. Gaynor, and S.D. Adams. (2006). "Undergraduate Research Communities: A Powerful Approach to Research Training." *Journal of College Science Teaching* July/August, 34-39.

[50] Kinkel, D. H. and S. E. Henke (2006). "Impact of undergraduate research on academic performance, educational planning, and career development." *Journal of Natural Resources and Life Sciences Education* 35: 194-201.

[51] Hunter, A-B., S. L. Laursen, and E. Seymour. (2007). "Becoming a scientist: The role of undergraduate research in students' cognitive, personal, and professional development." *Science Education* 91: 36-74.

[52] Russell, S. H., M.P. Hancock, and J. McCullough. (2007). "The pipeline: Benefits of undergraduate research experiences." *Science* 316: 548-549.

[53] Junge, B., C. Quiñones, J. Kakietek, D. Teodorescu, and P. Marsteller. (2010). "Promoting Undergraduate Interest, Preparedness, and Professional Pursuit in the Sciences: An Outcomes Evaluation of the SURE Program at Emory University." *CBE-Life Sciences Education* 9(2): 119-132.

[54] Espinosa. (2011). op. cit.

[55] The National Academies. (2010). op. cit.

[56] Conway, M. A. and S. J. Anderson. (1994). "The formation of flashbulb memories." *Memory & Cognition* 22: 326.

[57] Weigel, R. H. (1975). "The impact of cooperative learning experiences on cross-ethnic relations and attitudes." *Journal of Social Issues* 31: 219-244.

[58] Schwartz, D. L. and J.D. Bransford (1998). "A time for telling." *Cognition & Instruction* 16: 475-522.

[59] Springer, L., M. E. Stanne, and S.S. Donovan. (1999). "Effects of small-group learning on undergraduates in science, mathematics, engineering, and technology: A meta-analysis." *Review of Educational Research* 69(1): 21-51.

[60] Rivard, L. P. and S. B. Straw. (2000). "The effect of talk and writing on learning science: An exploratory study." *Science Education* 84: 566-593.

[61] Callender, A. A. and M. A. McDaniel. (2007). "The Benefits of Embedded Question Adjuncts for Low and High Structure Builders." *Journal of Educational Psychology* 99: 339-348.

[62] Chen, J. Y. (2011). "Problem-based learning: Developing resilience in nursing students." *Kaohsiung Journal of Medical Sciences* 27(6): 230-233.

[63] Morgan, R. L., J. E. Whorton, and C. Gunsalas. (2000). "A comparison of short term and long term retention: lecture combined with discussion versus cooperative learning." *Journal of Instructional Psychology* 27: 53-58.

[64] National Research Council. (2005). *How Students Learn: Science in the Classroom.* Washington, DC: National Academy Press.

[65] See The Posse Foundation: http://www.possefoundation.org/.

[66] Anderson W.A., U. Banerjee, C.L.Drennan, S.C. Elgin, I.R. Epstein, J. Handelsman, G.F. Hatfull, R. Losick, D.K. O'Dowd, B.M. Olivera, S.A. Strobel, G.C. Walker, and I.M.Warner. (2011). "Science education. Changing the culture of science education at research universities." *Science* 331(6014):152-3.

[67] See, for example, University of Wisconsin-Madison website for the Secretary of the Faculty: http://www.secfac.wisc.edu.

[68] Kezar, A. J. and P. D. Eckel. (2002). " The Effect of Institutional Culture on Change Strategies in Higher Education: Universal Principles or Culturally Responsive Concepts?" *Journal of Higher Education* 73(4): 435-460.

[69] National Research Council. (2005). *How Students Learn: Science in the Classroom.* Washington, DC: National Academy Press.

[70] J. Handelsman, et al.D. Ebert-May, R. Beichner, P. Bruns, A. Chang, R. DeHaan, J. Gentile, S. Lauffer, J. Stewart, S.Tilghman, W. Wood. (2004). "Scientific Teaching." *Science* 304(5670), 521–522.

[71] Ausubel, D. P. (2000). *The Acquisition and Retention of Knowledge: A Cognitive View.* Dordrecht, The Netherlands: Kluwer Academic Publishers.

[72] Peckham, J., P. Stephenson, J-Y Hervé, R. Hutt, and Miguel Encarnação. (2007). "Increasing student retention in computer science through research programs for undergraduates." *SIGCSE '07: Proceedings of the 38th SIGCSE Technical Symposium on Computer Science Education* 39(1): 124–128.

[73] McClanahan, E. B. and L. L. McClanahan. (2002). "Active Learning in a Non-Majors Biology Class." *College Teaching* 50: 92-96.

[74] Rivard, L. P., & Straw, S. B. (2000). "The effect of talk and writing on learning science: An exploratory study." *Science Education* 84 (5): 566-593.

[75] Schwartz, D. L., & Bransford, J. D. (1998). "A Time for telling." *Cognition and Instruction*, 16(4), 475-522.

[76] Schwartz, D.L., C. Chase, C. Chin, M. Oppezzo, H. Kwong, S. Okita, G. Biswas, R.D. Roscoe, H. Jeong, and J.D. Wagster. (2007). "Interactive metacognition: Monitoring and regulating a teachable agent." In D.J. Hacker, J. Dunlosky, and A.C. Graesser (Eds.), *Handbook of Metacognition and Education*. New York: Routledge.

[77] Roy, H. (2003). "Studio vs interactive lecture demonstration—effects on student learning." *Journal of College Biology Teaching* 29 (1): 3-6.

[78] Knight, J. K. and W.B. Wood, W. B. (2005). "Teaching more by lecturing less." *Cell Biology Education* 4, 298-310.

[79] Hake, R. R. (1998). "Interactive-engagement versus traditional methods: A six-thousand-student survey of mechanics test data for introductory physics courses." *American Journal of Physics* 66: 64-74.

[80] See also: http://cst.yale.edu/sites/default/files/Active%20learning%20research%20table%2012-27-11_0.pdf.

[81] The National Academies (2010). *Expanding Minority Participation: America's Science and Technology Talent at the Crossroads.* Washington, DC: National Academy Press.

[82] Pfund, C., S. Miller, K. Brenner, K., Bruns, P., Chang, A., Ebert-May, D., et al. (2009). "Summer Institute to Improve University Science Teaching." *Science* 324(5926): 470-471.

[83] See AAPT/APS/AAS Physics and Astronomy New Faculty Workshop: http://www.aapt.org /Conferences/newfaculty/nfw.cfm

[84] Henderson, C. (2008). *American Journal of Physics* 76 (2), 179-187. http://www.sigcse. org/events/symposia

[85] See The NPA Postdoctoral Core Competencies: http://www.nationalpostdoc.org /publications /competencies.

[86] See The CIRTL Network: http://www.cirtl.net. http://www.cirtl.net/.

[87] See ADVANCE program page: http://www.nsf.gov/crssprgm/advance/index.jsp.

[88] Stewart, A., J. Malley, and D. LaVaque-Manty (Eds.). (2007). *Transforming Science and Engineering: Advancing Academic Women*. Ann Arbor: University of Michigan Press.

[89] Sheridan, J.T., E. Fine, C.M. Pribbenow, J. Handelsman, and M. Carnes. (2010). "Searching for excellence & diversity: Increasing the hiring of women faculty at one academic medical center." *Academic Medicine* 85(6): 999-1007.

[90] Bilimore, D. (2011). *Gender Equity in Science and Engineering: Advancing Change in Higher Education.* New York, NY: Routledge Publishing Co.

[91] See Association of American Universities Undergraduate STEM Education Initiative: http://www.aau.edu/policy

[92] Alison Gopnik. (1999). *Small Wonders*. The New York Review of Books.

[93] Nagda, B. A., S. R. Gregerman, J. Jonides, W. von Hippel, and J.S. Lerner. (1998). "Undergraduate student-faculty research partnerships affect student retention." *The Review of Higher Education* 22: 55-72.

[94] Russell, S. H., Hancock, M.P., and J. McCullough. (2007). op. cit.

[95] Carter, F. D., M. Mandell, and K.I. Maton. (2009). "The Influence of On-Campus, Academic Year Undergraduate Research on STEM Ph.D. Outcomes: Evidence From the Meyerhoff Scholarship Program." *Educational Evaluation and Policy Analysis* 31(4): 441-462.

[96] Bednarksi, A.E., Elgin, S.C.R., and H.B. Pakrasi. (2005) "An Inquiry into Protein Structure and Genetic Disease: Introducing Undergraduates to Bioinformatics in a Large Introductory Course." *Cell Biology Education* Fall 2005.

[97] Pope, W. H., D. Jacobs-Sera, et al. (2011). "Expanding the diversity of mycobacteriophages: insights into genome architecture and evolution." *PLoS One* 6(1): e16329.

[98] The benchmarks specify the minimum scores needed on each ACT subject-area test to indicate that a student has a 50 percent chance of earning a grade of B or higher or about a 75 percent chance of earning a C or higher in a typical credit-bearing first-year college course in that subject area. ACT. (2011). *The Condition of College & Career Readiness* 2011. Iowa City, IA: ACT.

[99] Strong American Schools. (2008). *Diploma to Nowhere*. Washington, DC.

[100] Strong American Schools. (2008). op. cit.

[101] Bryk, A. S. and U. Triesman. (2010). "Make Math a Gateway, not a Gatekeeper." Washington, DC: *The Chronicle of Higher Education.*

[102] Lutzer, D. J., S. B. Rodi, E. E. Kirkman, and J.W. Maxwell. (2007). *Statistical Abstract of Undergraduate Programs in the Mathematical Science in the United States: Fall 2005 CBMS Survey*. Providence, RI: American Mathematical Society.

[103] National Mathematics Advisory Panel. (2008). *Foundations for Success.* Washington, DC: U.S. Department of Education.

[104] President's Council of Advisors on Science and Technology (PCAST). (2010) *Report to the President: Prepare and* Inspire *K-12 Education in Science, Technology, Engineering, and Math (STEM) for America's Future*. Washington, DC: PCAST. http://www.whitehouse.gov /sites/default/files/microsites/ostp/pcast-stemed-report.pdf.

[105] Bettinger E. and B. Long, B. (2009). "Addressing the Needs of Underprepared Students in Higher Education: Does College Remediation Work?" *Journal of Human Resources* 44(3): 736–771.

[106] See Carnagie Melon University's website: http://www.cmu.edu/enrollment/summerprograms fordiversity/ sams.html.

[107] See California State Summer School for Mathematics and Science: http://www.ucop.edu /cosmos/.

[108] Chen, G. (2009) T*he Minority Report: How Minority Students are Really Faring at Community Colleges.* Community College Review: http://www.community collegereview. com/articles/202.

[109] Phillippe, K.A. and L. Gonzalez Sullivan. (2005). *National Profile of Community Colleges: Trends & Statistics.* American Association of Community Colleges. Washington DC: Community College Press.

[110] Reyes, M.-E. (2011). "Unique Challenges for Women of Color in STEM Transferring from Community Colleges to Universities." *Harvard Educational Review Summer,* 241-263.

[111] Cullinane, J. and L. H. Leegwater. (2009). *Diversifying the STEM Pipeline: The Model Replication Institutions Program.* Washington, DC: Institute for Higher Education Policy.

[112] Southern Education Foundation. (2005). *Igniting Potential: Historically Black Colleges and Universities and Science, Technology, Engineering and Mathematics*. Atlanta, GA.

[113] Brown, S. (2009). "Making the Next Generation our Greatest Resource" in *Latinos and the Nation's Future*, H. Cisneros and J. Rosales (eds.) Houston, TX: Arte Publico Press.

[114] See White House Initiative on Historically Black Colleges and Universities: http://www2.ed. gov/about/inits/list/whhbcu/edlite-index.html.

[115] Hess, F.M., A.P. Kelly, and O. Meeks. (2011). *The Case for Being Bold: A New Agenda for Business in Improving STEM Education*. Washington, DC: Institute for a Competitive Workforce.

[116] See President's Council on Jobs and Competiveness website: http://www.whitehouse.gov /administration/advisory-boards/jobs-council.

[117] See Jobs Council Internship Commitment Announcement. http://energy.

[118] See http://www.whitehouse.gov/administration/eop/ostp/pcast/amp.

[119] President's Council of Advisors on Science and Technology (PCAST) (2011). R*eport to the President on Ensuring American Leadership in Advanced Manufacturing*. Washington, DC: PCAST. http://www. whitehouse.gov/sites/default/files/ microsites/ostp/pcast-advanced-manufacturing-june2011.pdf.

[120] See IBM's 'IBM "Transition to Teaching"Teaching' Program: http://www.ibm.com/ibm /responsibility/teaching.shtml.

[121] Bresnahan, Timothy F., Erik Brynjolfsson, and Lorin M. Hitt. (2002). "Information Technology, Workplace Organization and the Demand for Skilled Labor: Firm-Level Evidence." *Quarterly Journal of Economics*, 117, 339-376. For a contrasting view, see: Michael J. Handel. (2005). "Worker Skills and Job Requirements: Is There a Mismatch?" Washington, DC: Economic Policy Institute.

[122] Carnevale, Anthony, Nichole Smith, and Jeff Strohl. (2010). "Help Wanted: Projections of Jobs and Education Requirements through 2018." Washington, DC: Georgetown University Center on Education and the Workforce.

[123] The benchmarks specify the minimum scores needed on each ACT subject-area test to indicate that a student has a 50 percent chance of earning a grade of B or higher or about a 75 percent chance of earning a C or higher in a typical credit-bearing first-year college course in that subject area. ACT. (2011). *The Condition of College & Career Readiness 2011*. Iowa City, IA: ACT.

[124] Strong American Schools. (2008). *Diploma to Nowhere.* Washington, DC.

[125] National Research Council. (2005). *How Students Learn: Science in the Classroom*. Washington, DC: National Academy Press.

[126] Rivard, L.P. and S.B. Straw. (2000). "The effect of talk and writing on learning science: An exploratory study." *Science Education* 84: 566-593.

[127] Kornell, N. and H.S. Terrace. (2007). "The generation effect in monkeys." *Psychological Science* 18(8): 682-685.

[128] Woods, D., A. Hrymak, R. Marshall, P. Wood, C. Crowe, and T. Hoffman (1997). "Developing problem solving skills: The McMaster Problem Solving Program." *Journal of Engineering Education,* April, 75-91.

[129] Deslauriers, L., E. Schelew, and C. Wieman (2011). "Improved learning in a large-enrollment physics class." *Science* 332(6031): 862-4.

[130] Smith, M.K, W.B. Wood, K. Krauter, and J.K Knight. (2011). "Combining peer discussion with instructor explanation increases student learning from in-class concept questions." *CBE Life Sciences Education* 10: 55-63.

[131] A general survey of promising practices in undergraduate STEM education, with an emphasis on the extent to which the practices have been validated by research, is the white paper "Promising Practices in Undergraduate STEM Education" (2008) by J. E. Froyd. Available at http://www7.nationalacademies.org/bose /Froyd_Promising_Practices_ Commissioned Paper.pdf. See also Baldwin, Roger G., ed. (2009). *Improving the Climate for Undergraduate Teaching and Learning in STEM Fields*. San Francisco: Jossey-Bass.

[132] Hake, R.R. (1998). "Interactive engagement versus traditional methods: A six-thousand-student survey of mechanics test data for introductory physics courses." *American Journal of Physics* 66(1): 64.

[133] Springer, L., M.E. Stanne, and S.S. Donovan (1999). "Effects of small-group learning on undergraduates in science, mathematics, engineering, and technology: A meta-analysis." *Review of Educational* Research 69(1): 21-51.

[134] Bowen, C.W. (2000). "A quantitative literature review of cooperative learning effects on high school and college chemistry achievement." *Journal of Chemical Education* 77(1): 116.

[135] Ibid.

[136] Hoffman, K., M. Hosokawa, R. Blake, L. Headrick, and G. Johnson. (2006). "Problem-based learning outcomes: ten years of experience at the University-Columbia School of Medicine." *Academic Medicine: Journal of the Association of American Medical Colleges* 81(7): 617-25.

[137] Ibid.

[138] Smith, M., W. Wood, W. Adams, C. Wieman, J. Knight, N. Guild. (2009). "Why peer discussion improves student performance in class." *Science* 323: 122-124.

[139] Schwartz, D.L. and J.D. Bransford. (1998). "A time for telling." *Cognition and Instruction* 16(4): 475-522.

[140] Buck, J.R. and K.E. Wage. (2005). "Active and cooperative learning in signal processing courses." *IEEE Signal Processing Magazine* 22(2): 76-81.

[141] Capon, N. and D. Kuhn (2004). "What's so good about problem-based learning." *Cognition and Instruction* 22(1): 61-79.

[142] Almer, E., Jones, K., and Moeckel, C. (1998). The impact of one-minute papers on learning in an introductory accounting course. Issues in Accounting Education 13(3): 485-495.

[143] McDaniel, M., J. Anderson, M. Derbish, and N. Morrisette (2007). "Testing the testing effect in the classroom." *European Journal of Cognitive Psychology* 19(4): 494-513.

[144] Brickman, P., C. Gormally, N. Armstrong, and B. Hallar. (2009). "Effects of Inquiry-based Learning on Students' Science Literacy Skills and Confidence." *International Journal for the Scholarship of Teaching and Learning.*

[145] Nagda, B. A., S.R. Gregerman, J. Jonides, W. von Hippel, and J.S. Lerner (1998). "Undergraduate student-faculty research partnerships affect student retention." *Review of Higher Education* 22(1), 55-72.

[146] Russell, S.H., M.P. Hancock, and J. McCullough. (2007). "Benefits of undergraduate research experience." *Science* 316: 548-9.

[147] See Louisiana Science, Technology, Engineering and Mathematics Research Scholars Program Website: http://www.lsu.edu/lastem/files/LA-STEM%20flyer%20for%20LSU z%20and%20Transfer%20Students%202011.pdf.

[148] Cullinane, J. and L.H. Leegwater (2009). *Diversifying the STEM Pipeline: The Model Replication Institutions Program*. Washington, DC: Institute for Higher Education Policy.

In: Strategies for Effective Science, Technology, Engineering... ISBN: 978-1-62081-523-6
Editors: Valerian I. Bell and Shioban E. Gonzalez

Chapter 2

PRESIDENTIAL COMMITMENTS IN SUPPORT OF PCAST RECOMMENDATIONS ON SCIENCE, TECHNOLOGY, ENGINEERING, AND MATHEMATICS EDUCATION*

Executive Office of the President of the United States

A number of commitments announced by President Obama today in conjunction with the White House Science Fair respond directly to recommendations in two reports by the President's Council of Advisors on Science and Technology.

In September 2010, PCAST released *Prepare and Inspire: K-12 Education in Science, Technology, Engineering, and Math (STEM) for America's Future*, which, among other recommendations, called for a major push to recruit and train 100,000 great STEM teachers over the next decade able to prepare and inspire students. In support of that goal, the President announced today:

- **A new $80 million investment to help prepare effective STEM teachers**: The President's upcoming budget will request $80 million for a new competition by the Department of Education to support effective STEM teacher preparation programs, such as those that allow students to simultaneously earn both a STEM degree and a teaching certificate, and provide undergraduates with early and intensive experiences in the classroom honing their skills.
- **A new $22 million investment from the philanthropic and private sector to complement the Administration's efforts**: After the President issued his call to action to recruit and prepare 100,000 effective STEM teachers, over 115 organizations, led by Carnegie Corporation of New York and Opportunity Equation, came together to form a coalition called "100Kin10" to help reach the President's goal. Today, 14 of those organizations – including Carnegie, Google, the S.D.

* This is an edited, reformatted and augmented version of Executive Office of the President of the United States, dated February 7, 2012.

Bechtel, Jr., Bill & Melinda Gates, Freeport McMoran, and Michael and Susan Dell Foundations – are announcing a $22 million fund to invest in STEM teacher preparation. In addition, several other 100Kin10 partners are making over 100 individual commitments, such as:

- **Citizen Schools** will bring 10,000 additional STEM professionals into classrooms to teach part-time over the next 10 years;
- **National Math and Science Initiative** will prepare 4,000 new STEM teachers from 31 UTeach sites by 2015;
- **Teach for America** will recruit 11,000 STEM Corps members by 2015 and connect other qualified applicants to additional STEM teaching opportunities;
- **Donors Choose** will inspire 50,000 citizens to sponsor projects in math and science classrooms over the next two years, delivering $15M in critical classroom resources and helping 600,000 students nationwide;
- **Google** will share its talent management practices to help find, grow, and retain outstanding STEM teachers by partnering with districts and organizations for comprehensive reform and hosting talent academies with administrators and decision-makers; and,
- **University of Chicago** will create a framework for organizing the learning that results from "100Kin10" investments and coordinate research among partners on key questions about STEM teacher recruitment, preparation, induction, and development.

A complete list of partners, their commitments, and general information about 100Kin10 is available at their website: www.100Kin10.org.

- **New policies and investments to recruit, support, retain and reward excellent STEM teachers:** To improve the teaching and learning of STEM and encourage our best STEM teachers to stay in the profession, we must implement a system that recognizes and rewards teacher excellence. That's why, this year, the Department of Education will devote a portion of its upcoming $300 million Teacher Incentive Fund competition to support state and local efforts to improve compensation, evaluation, and professional development systems for STEM educators. In addition, the Department of Education will provide new incentives to improve the quality of teacher preparation programs by targeting TEACH Grants to students attending top-tier schools, and focusing on a smaller number of more meaningful outcome indicators about their quality and impact on teacher performance. Concurrently, the National Science Foundation will continue to emphasize the quality of teacher preparation programs and plans for innovation in its Robert Noyce Scholarship program.

Today, the President announced his support for the goal of producing, over the next decade, 1 million more STEM professionals than are projected to graduate at current rates—a goal at the core of PCAST's newest report, *Engage to Excel: Producing One Million Additional College Graduates with Degrees in Science, Technology, Engineering, and Mathematics*, released today. In support of that goal, the President committed to:

- **A priority on undergraduate STEM education reform in the President's upcoming budget:** The President announced more than $100 million in investments by the National Science Foundation to improve undergraduate STEM education practices through its programs such as Widening Implementation and Demonstration of Evidence-based Reforms (WIDER), Transforming Undergraduate Education in Science, Technology, Engineering and Mathematics (TUES), and programs that impact community colleges and minority-serving institutions. This will support the development, identification, and scale-up of educational practices that increase the number of STEM graduates and the quality of their preparation. In addition, the Department of Education's proposed First in the World competition will include a STEM priority.
- **A new K-16 education initiative jointly administered by the Department of Education and the National Science Foundation**: To support comprehensive reform efforts in K-16 education, the President's FY 2013 budget will fund a jointly administered initiative to improve mathematics education, with $30 million from the Department of Education and $30 million from the National Science Foundation. This initiative will develop, validate, and scale up evidence-based approaches to improve student learning at the K-12 and undergraduate levels through a "tiered-evidence framework" to maximize the impact of mathematics education investments.

In: Strategies for Effective Science, Technology, Engineering... ISBN: 978-1-62081-523-6
Editors: Valerian I. Bell and Shioban E. Gonzalez

Chapter 3

SCIENCE, TECHNOLOGY, ENGINEERING, AND MATHEMATICS EDUCATION: STRATEGIC PLANNING NEEDED TO BETTER MANAGE OVERLAPPING PROGRAMS ACROSS MULTIPLE AGENCIES*

United States Government Accountability Office

WHY GAO DID THIS STUDY

Science, technology, engineering, and mathematics (STEM) education programs help to enhance the nation's global competitiveness. Many federal agencies have been involved in administering these programs. Concerns have been raised about the overall effectiveness and efficiency of STEM education programs.

GAO examined (1) the number of federal agencies and programs that provided funding for STEM education programs in fiscal year 2010; (2) the extent to which STEM education programs have similar objectives, serve similar target groups, and provide similar types of services, and, if necessary, what opportunities exist to increase coordination; and (3) the extent to which STEM education programs measured effectiveness. To answer these questions, GAO reviewed relevant federal laws, regulations, and plans; surveyed federal STEM education programs; analyzed programs' STEM evaluations; and interviewed relevant federal officials. An electronic supplement— GAO-12-110SP—provides survey results.

WHAT GAO RECOMMENDS

GAO recommends that as OSTP leads the governmentwide STEM education strategic planning effort, it should work with agencies to better align their activities with a governmentwide strategy, develop a plan for sustained coordination, identify programs for

* This is an edited, reformatted and augmented version of United States Government Accountability Office, Report to Congressional Requesters, Publication No. GAO-12-108, dated January 2012.

potential consolidation or elimination, and assist agencies in determining how to better evaluate their programs. OSTP provided technical comments that we incorporated as appropriate. OMB had no concerns with the report.

WHAT GAO FOUND

In fiscal year 2010, 13 federal agencies invested over $3 billion in 209 programs designed to increase knowledge of STEM fields and attainment of STEM degrees. The number of programs within agencies ranged from 3 to 46, with the Departments of Health and Human Services and Energy and the National Science Foundation administering more than half of these programs. Almost a third of the programs had obligations of $1 million or less, while some had obligations of over $100 million. Beyond programs specifically focused on STEM education, agencies funded other broad efforts that contributed to enhancing STEM education. Eighty-three percent of the programs GAO identified overlapped to some degree with at least 1 other program in that they offered similar services to similar target groups in similar STEM fields to achieve similar objectives. Many programs have a broad scope—serving multiple target groups with multiple services. However, even when programs overlap, the services they provide and the populations they serve may differ in meaningful ways and would therefore not necessarily be duplicative. Nonetheless, the programs are similar enough that they need to be well coordinated and guided by a robust strategic plan. Currently, though, less than half of the programs GAO surveyed indicated that they coordinated with other agencies that administer similar STEM education programs. Current efforts to inventory federal STEM education activities and develop a 5-year strategic plan present an opportunity to enhance coordination, align governmentwide efforts, and improve efficiency of limited resources by identifying opportunities for program consolidation and reducing administrative costs. Agencies' limited use of performance measures and evaluations may hamper their ability to assess the effectiveness of their individual programs as well as the overall STEM education effort. Specifically, program officials varied in their ability to provide reliable output measures—for example, the number of students, teachers, or institutions directly served by their program. Further, most agencies did not use outcomes measures in a way that is clearly reflected in their performance planning documents. This may hinder decision makers' ability to assess how agencies' STEM education efforts contribute to agencywide performance goals and the overall federal STEM effort. In addition, a majority of programs did not conduct comprehensive evaluations since 2005 to assess effectiveness, and the evaluations GAO reviewed did not always align with program objectives. Finally, GAO found that completed STEM education evaluation results had not always been disseminated in a fashion that facilitated knowledge sharing between both practitioners and researchers.

ABBREVIATIONS

ACC	Academic Competitiveness Council
COMPETES	America COMPETES Act
DHS	Department of Homeland Security

DOD	Department of Defense
DOT	Department of Transportation
EPA	Environmental Protection Agency
GPRA	Government Performance and Results Act
HHS	Department of Health and Human Services
HIS	Indian Health Service
K-12	Kindergarten-12th grade
NASA	National Aeronautics and Space Administration
CoSTEM	Committee on Science, Technology, Engineering, and Math Education
NIH	National Institutes of Health
NRC	Nuclear Regulatory Commission
NSF	National Science Foundation
NSTC	National Science and Technology Council
OMB	Office of Management and Budget
OSTP	Office of Science and Technology Policy
PCAST	President's Council of Advisors on Science and Technology
SMD	NASA Science Mission Directorate
STEM	Science, Technology, Engineering, and Mathematics
USDA	Department of Agriculture

January 20, 2012
The Honorable John Kline
Chairman,
Committee on Education and the Workforce
House of Representatives

The Honorable Duncan D. Hunter
Chairman,
Subcommittee on Early Childhood, Elementary,
and Secondary Education, Committee on Education and the Workforce
House of Representatives

Federally funded science, technology, engineering, and mathematics (STEM) education programs can serve an important role both by helping to prepare students and teachers for careers in STEM fields and by enhancing the nation's global competitiveness. In this effort, many federal agencies administer STEM education programs. In addition to the federal effort, state and local governments, universities and colleges, and the private sector have also developed programs that provide opportunities for students to pursue STEM education and occupations. Nonetheless, research continues to show that the United States lacks a strong pipeline of future workers in STEM fields and that U.S. students continue to lag behind students in other highly technological nations in mathematics and science achievement. Over the decades, Congress and the executive branch have continued to create new STEM education programs, even though, as we reported in 2005, there has been a general lack of assessment of how well STEM programs are working.[1] A little more than a year after our report was issued, the Academic Competitiveness Council (ACC)—headed by the Department of Education—issued a report that outlined areas of potential overlap and recommended areas for better coordination and evaluation of STEM education programs.[2] In this context, we were asked to examine the delivery and effectiveness of STEM education programs. Specifically, our objectives were to determine (1) the number of federal agencies and programs that provided funding for STEM education programs in fiscal year 2010; (2) the extent to which these STEM programs have similar objectives, serve similar target groups, and provide similar types of services and, if necessary, what opportunities exist to increase coordination; and (3) the extent to which federal STEM education programs have measured their effectiveness. To address our objectives, we collected and analyzed information through several methods. We reviewed relevant federal laws and regulations as well as previous GAO work on overlap, duplication, and fragmentation. We interviewed officials from the Office of Management and Budget (OMB) and the Office of Science and Technology Policy (OSTP), and officials from other federal agencies that administer STEM education programs. We reviewed relevant literature and past reports that catalog and assess the federal investment in STEM education. To gather information on federal STEM education programs and to assess the level of fragmentation, overlap, and potential duplication, we surveyed over 200 programs across 13 agencies that met our definition of a STEM education program, asking questions about program objectives, target populations, services provided, interagency coordination, outcome measures and evaluations, and funding.[3] Our web-based survey, which was administered between May 2011 and August 2011 to federal agency program officials, achieved a 100 percent response rate. To assess the reliability of data provided in our survey, we incorporated questions about the reliability of the programs' data systems, reviewed documentation for a sample of selected questions, conducted internal reliability checks, and conducted follow-up as necessary. While we did not verify all responses, we determined that the data used in our report are sufficiently reliable for our purpose. To gather additional perspectives about federal STEM education programs, we attended several STEM education conferences. To gather information on program effectiveness, we reviewed evaluations provided by program officials as well as agencies' annual performance plans and reports. For more information on our scope and methodology, see appendix I. The STEM survey and selected results can be found in GAO-12-110SP, an e-supplement that is a companion to this report. We conducted this performance audit from February 2011 through January 2012, in accordance with generally accepted government auditing standards. Those standards require that we plan and perform the audit to obtain sufficient, appropriate evidence to provide a reasonable basis for our findings and conclusions based on our audit objectives. We believe that the evidence obtained provides a reasonable basis for our findings and conclusions based on our audit objectives.

BACKGROUND

Past Efforts to Assess Federal STEM Education Efforts

In 2005, we reported that 207 federal STEM education programs across 13 different agencies spent $2.8 billion in federal funds in fiscal year 2004.[4] We noted that before increasing investment in STEM education, it is important to know the extent to which existing STEM education programs are appropriately targeted and whether or not they are making the best use of available federal resources. Additionally, information about the effectiveness of these programs could help guide policymakers and program managers.

Table 1. Agencies Administering STEM Education Programs

Agency	Mission
Department of Agriculture (USDA)	To provide leadership on food, agriculture, natural resources, and related issues based on sound public policy, the best available science, and efficient management
Department of Commerce (Commerce)	To promote job creation, economic growth, sustainable development, and improved standards of living for all Americans by working in partnership with businesses, universities, communities, and our nation's workers
Department of Defense (DOD)	To provide the military forces needed to deter war and to protect the security of our country
Department of Education (Education)	To promote student achievement and preparation for global competitiveness by fostering educational excellence and ensuring equal access
Department of Energy (Energy)	To ensure America's security and prosperity by addressing its energy, environmental and nuclear challenges through transformative science and technology solutions
Department of Health and Human Services (HHS)	To enhance the health and well-being of Americans by providing for effective health and human services and by fostering sound, sustained advances in the sciences underlying medicine, public health, and social services
Department of Homeland Security (DHS)	To ensure a homeland that is safe, secure, and resilient against terrorism and other hazards
Department of the Interior (Interior)	To protect and manage the nation's natural resources and cultural heritage; to provide scientific and other information about those resources; and to honor its trust responsibilities or special commitments to American Indians, Alaska Natives, and affiliated island communities
Department of Transportation (DOT)	To ensure a fast, safe, efficient, accessible and convenient transportation system that meets our vital national interests and enhances the quality of life of the American people, today and into the future
Environmental Protection Agency (EPA)	To protect human health and the environment
National Aeronautics and Space Administration (NASA)	To drive advances in science, technology, and exploration to enhance knowledge, education, innovation, economic vitality, and stewardship of Earth
National Science Foundation (NSF)	To promote the progress of science; to advance the national health, prosperity, and welfare; to secure the national defense; to support science and engineering education, from prekindergarten through graduate school and beyond, among other things
Nuclear Regulatory Commission (NRC)	To ensure the adequate protection of public health, safety, and the environment while promoting the common defense and security

Source: GAO review of agencies' websites and strategic plans.

Since then, several other efforts have been conducted to identify federal STEM programs and provide recommendations to improve both coordination and program evaluation as well as reduce potential duplication. For example, in 2006, ACC, led by the Department of Education, created an inventory and assessed the effectiveness of federal STEM programs. ACC recommended further coordination among federal agencies administering STEM programs, states, and local school districts. In addition, ACC recommended that agencies adjust program designs and operations so that programs can be assessed and measurable results can be achieved and that funding for federal STEM education programs should not be increased unless a plan for rigorous, independent evaluation is in place.

In 2010, the President's Council of Advisors on Science and Technology (PCAST), an advisory group of the nation's leading scientists and engineers housed in OSTP, published a report in response to the President's request to develop specific recommendations concerning the most important actions that the administration should take to ensure that the United States is a leader in STEM education in the coming decades.[5] PCAST found that approaches to Kindergarten–12th grade (K-12) STEM education across agencies emerged largely without a coherent vision or careful oversight of goals and outcomes.

PCAST also found that relatively little funding was targeted at efforts with the potential to transform STEM education, too little attention was paid to replication efforts to disseminate proven programs widely, and too little capacity at key agencies was devoted to strategy and coordination.

Our past effort to inventory STEM education programs identified a multitude of agencies that administer such programs. The primary missions of these agencies vary, but most often, they are to promote and enhance an area that is related to a STEM field or enhance general education. See table 1 for relevant agencies and their missions.

As part of this effort, we also identified the role that the National Science and Technology Council (NSTC), a component of OSTP, plays in coordinating STEM education programs. NSTC was established in 1993 and is the principal means for the administration to coordinate science and technology with the federal government's larger research and development effort.[6] NSTC is made up of the Vice President, the Director of the Office of Science and Technology Policy, and officials from other executive branch agencies with significant science and technology responsibilities. One objective of NSTC is to establish clear national goals for federal science and technology investments in areas ranging from information technologies and health research to improving transportation systems and strengthening fundamental research. NSTC is responsible for preparing research and development strategies that are coordinated across federal agencies in order to accomplish these multiple national goals.

Federal Legislation

STEM education programs have been created in two ways—by Congress directly in legislation or through agencies' broad statutory authority to carry out their missions.[7] The Higher Education Opportunity Act,[8] the No Child Left Behind Act of 2001,[9] and the National Science Foundation Act of 1950[10] created programs at the Department of Education and the National Science Foundation (NSF)—two key agencies that administer many STEM education programs. In addition, since our 2005 review of STEM education programs, Congress has also passed legislation to examine the overall federal effort to improve STEM education. For example, the Deficit Reduction Act of 2005 established ACC.[11] ACC, consisted of officials from the Department of Education and other federal agencies with responsibility for managing mathematics and science education programs and was mandated to (1) identify all federal programs with a mathematics or science education focus, (2) identify the target populations being served by such programs, (3) determine the effectiveness of such programs, (4) identify areas of overlap or duplication in such programs, and (5) recommend processes to integrate and coordinate such programs. While various pieces of legislation directly created some STEM education programs, agencies reported using their broad statutory authority to create many programs as well. For example, according to agency officials, NSF created 25 of its 37 programs and the Department of Health and Human Services (HHS) created 40 of its 46 programs in this manner.

More recently, the America COMPETES Act (COMPETES), enacted in 2007, authorized several programs to promote STEM education. [12] In December 2010, Congress reauthorized COMPETES.[13] The reauthorization approved new funding for some STEM education programs and made substantive changes to others by reducing certain nonfederal matching requirements. Additionally, it repealed many of the programs that went unfunded following the original COMPETES passage.

The COMPETES reauthorization also sought to address coordination and oversight issues, including those associated with the coordination and potential duplication of federal STEM education efforts. Specifically, Congress required the Director of OSTP to establish a committee under NSTC to inventory, review, and coordinate federal STEM education programs.[14] Congress also directed this NSTC committee to specify and prioritize annual and long-term objectives for STEM education, and to ensure that federal efforts do not duplicate each other, among other things. NSTC is required to report to Congress annually.

Beyond STEM-specific efforts, the federal government as a whole is seeking to identify programmatic areas that could be better tracked and coordinated. One such effort revolves around the Government Performance and Results Act (GPRA) Modernization Act of 2010. [15] The GPRA Modernization Act established a new framework aimed at taking a more crosscutting and integrated approach to focusing on results and improving government performance. It requires OMB, in coordination with agencies,[16] to develop—at least every 4 years—long-term priority goals, including outcome-oriented goals covering a limited number of crosscutting policy areas. On an annual basis, OMB is to provide information on how these long-term crosscutting goals will be achieved. This approach could provide a basis for more fully integrating a wide array of federal activities as well as a cohesive perspective on the longterm goals of the federal government.

GAO's Work on Fragmentation, Overlap, and Duplication

In 2010, Congress directed GAO to conduct routine investigations to identify programs, agencies, offices, and initiatives with duplicative goals and activities within departments and governmentwide and report annually to Congress.[17] In March 2011, GAO issued its first annual report to Congress in response to this requirement.[18] In that report, we identified 81 areas for consideration—34 areas of fragmentation, overlap, and potential duplication and 47 additional areas—where agencies or Congress may wish to consider taking action in an effort to reduce the cost of government operations or enhance revenue collections. Using the framework established in the March 2011 GAO report, we examine the extent to which federal STEM education programs are fragmented, overlapping, and duplicative.

For the purposes of this report, the key terms are defined as follows:

- Fragmentation occurs when more than one federal agency (or more than one organization within an agency) is involved in the same broad area of national need.
- Overlap occurs when multiple programs offer similar services to similar target groups in similar STEM fields to achieve similar objectives.

- Duplication occurs when multiple programs offer the same services to the same target beneficiaries in the same STEM fields.

THIRTEEN FEDERAL AGENCIES ADMINISTERED OVER 200 STEM EDUCATION PROGRAMS WITH OVER $3 BILLION IN OBLIGATED FUNDS

Thirteen agencies administered 209 STEM education programs in fiscal year 2010.[19] (See appendix I for our definition of a STEM education program.) Agencies reported that they developed the majority (130) of these programs through their general statutory authority and that Congress specifically directed agencies to create 59 of these programs.[20] The number of programs each agency administered ranged from 3 to 46 with three agencies—HHS, the Department of Energy, and NSF— administering more than half of all programs—112 of 209. Figure 1 provides a summary of the number of programs by agency, and appendix II contains a list of the 209 STEM education programs and reported obligations for fiscal year 2010.

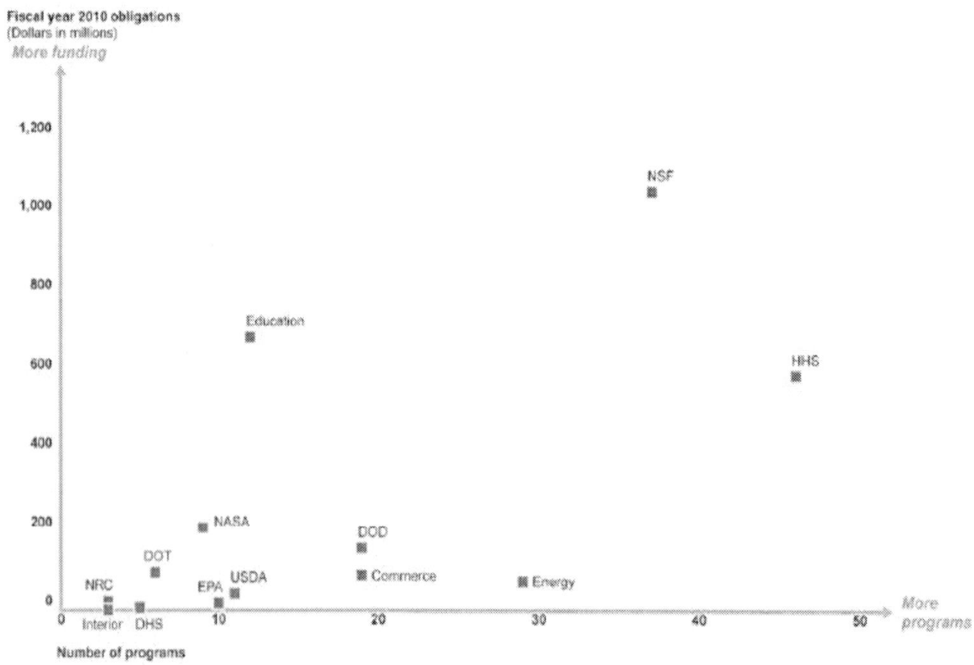

Figure 1. Number of STEM Education Programs Reported by Agency.

Having multiple agencies, with varying expertise, involved in delivering STEM education can be advantageous. One such advantage is that agencies may be better able to tailor programs to suit their specific missions and needs. For example, Energy officials said that their efforts to support students in pursuing a STEM course of study are related to Energy's mission and work in their labs and can be a way to attract new employees to their workforce. However, this could also make it challenging to develop a coherent federal approach to educating STEM students and creating a workforce with STEM skills. Having multiple agencies involved in the delivery of STEM education could also make it challenging to identify gaps and allocate resources across the federal government.

Agencies obligated over $3 billion to STEM education programs in fiscal year 2010.[21] Individual program obligations ranged from $15,000 to hundreds of millions of dollars. NSF and the Department of Education programs account for over half of this funding. Almost a third of the programs had obligations of $1 million or less, with 5 programs having obligations of more than $100 million each. See figure 2 for program obligation ranges.

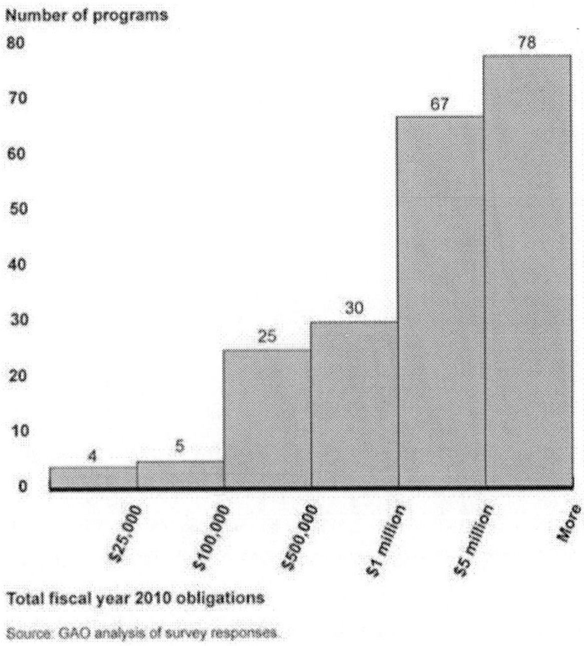

Source: GAO analysis of survey responses.

Figure 2. Number of STEM Education Programs by Range of Obligations, Fiscal Year 2010.

Agencies carried out other activities that did not fit our definition of a STEM education program because STEM education was their secondary or tertiary objective, rather than their primary objective. These efforts include broad-based programs with STEM components, programs that enhance the general public's knowledge of STEM, and research programs that may hire students.[22] Selected examples of agencies' efforts as reported to us by agency officials include the following:

Broad-Based Programs That Include STEM Components

- Several of the Department of Education's programs have STEM components. For example, Title I of the Elementary and Secondary Education Act of 1965, as amended, includes funding for the assessment of math for primary and secondary students, putting a renewed focus on educational attainment in these areas. In addition, the Race to the Top Fund, a competitive grant program, includes bonus points for states that report they will include in their grant activity, efforts to enhance STEM education.

- The Department of Transportation's State Maritime Academy program supports maritime training and education programs in an effort to improve the quality of the U.S. maritime industry with a secondary objective to encourage students to pursue careers in STEM fields that can contribute to the maritime industry.

Programs to Educate the General Public

- The National Institutes of Health's (NIH) Science Education Drug Abuse Partnership Award provides support for the formation of partnerships among scientists and educators, media experts, community leaders, and other interested organizations for the development and evaluation of programs and materials that will enhance knowledge and understanding of science related to drug abuse. The intended focus is on topics not well addressed in existing efforts by educational, community, or media activities.

Research Programs That Include Internships or Assistantships

- Energy's national laboratories, most of which are managed by contractors and engage in research activities on behalf of multiple federal agencies, sometimes partner with universities and offer students research opportunities in various disciplines, such as science and technology. The primary focus of these laboratories is on research and development, which is determined by the funding institution, and there is not always a requirement that they hire students. When research programs do hire students, this can enhance students' education and interest in STEM.
- The Department of Defense has several programs with a primary objective to further research on a specific STEM topic. For example, it has programs that fund university faculty to conduct research on STEM topics and who may hire students to assist with research.
- The Department of Homeland Security receives funding for technological research in areas that support its mission, and a portion of this may go to student research activities such as hiring a student for the summer or for several weeks to assist with the research.

Nonmonetary Partnerships with Schools or through Private Partnerships

- The Department of the Interior participates in the GeoFORCE program—a precollege program that provides hands-on science learning experiences for middle and high school students (primarily underserved minorities)—which is mostly funded by private donations and the University of Texas.[23]
- The Environmental Protection Agency has a cooperative agreement with the Hispanic Association of Colleges and Universities that is intended to increase the

diversity of students going into science and technology careers. The agreement includes activities such as EPA staff participation in lectures, conferences, and other events, as well as EPA staff members serving as mentors or coaches, among other things.

Dedicated Funds for Education Programs

- NASA's Science Mission Directorate (SMD) requires each of its missions to fund SMD-related education and public outreach using a small percentage of the research and development program costs, but these funds are not specifically for STEM education.

MOST STEM PROGRAMS OVERLAPPED TO SOME DEGREE, HIGHLIGHTING THE NEED FOR IMPROVED COORDINATION AND PLANNING

Most Programs Overlapped to Some Degree in Their Primary Objectives, Target Groups, and Services Provided

As figure 3 illustrates, in fiscal year 2010, 83 percent of STEM education programs overlapped to some degree with another program in that they offered at least one similar service to at least one similar target group in at least one similar STEM field to achieve at least one similar objective. These programs ranged from being narrowly focused on a specific group or field of study to offering a range of services to students and teachers across STEM fields. This complicated patchwork of overlapping programs has largely resulted from federal efforts to both create and expand programs across many agencies in an effort to improve STEM education and increase the number of students going into STEM fields. Program officials reported that approximately one-third of STEM education programs funded in fiscal year 2010 were first funded between 2005 and 2010. Indeed, the creation of new programs during that time frame may have contributed to overlap and, ultimately, to inefficiencies in how STEM programs across the federal government are focused and delivered. Overlap among STEM education programs is not new. In 2007, ACC identified extensive overlap among STEM education programs, and, in 2009, we identified overlap among teacher quality programs, which include several programs focused on STEM education.

Similar Target Groups

Many programs provided services to similar target groups, such as K-12 students, postsecondary students, K-12 teachers, and college faculty and staff. The vast majority of programs (170) served postsecondary students. Ninety-five programs served college faculty and staff, 75 programs served K-12 students, and 70 programs served K-12 teachers. In addition, many programs served multiple target groups. In fact, as figure 4 illustrates, 177 programs were primarily intended to serve two or more target groups.

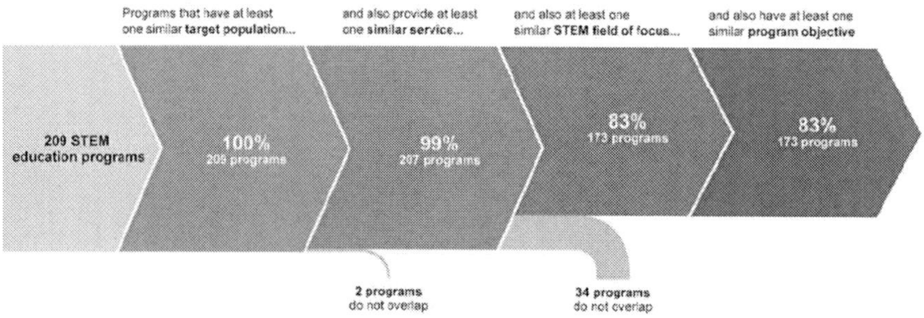

Source: GAO analysis of survey responses.

Figure 3. Overlapping Federal STEM Education Programs.

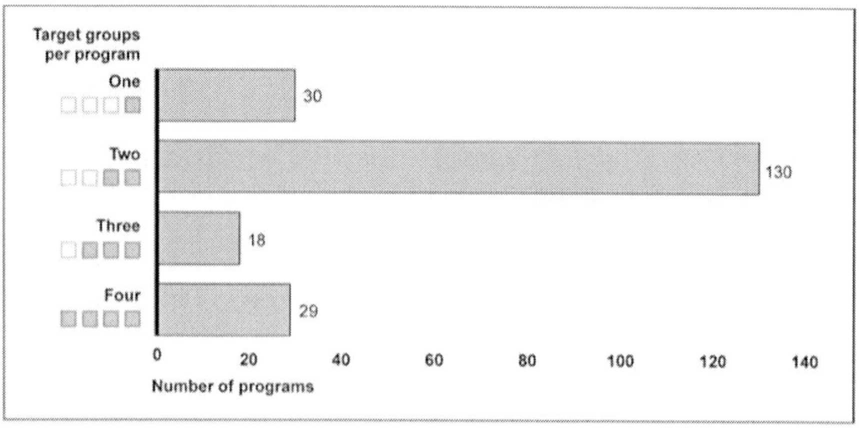

Source: GAO analysis of survey responses.
Note: Two programs indicated they did not serve any of the target groups listed in our survey.

Figure 4. Number of Target Groups per Federal STEM Education Program.

Similar Services Provided

As figure 5 illustrates, we also found many STEM programs providing similar services.

- To support students, 167 different programs provided research opportunities, internships, mentorships, or career guidance. In addition, 144 programs provided short-term experiential learning opportunities and 127 long-term experiential learning opportunities. Short-term experiential learning activities include field trips, guest speakers, workshops, and summer camps. Long-term experiential learning activities last a semester in length or longer. Furthermore, 137 programs provided outreach and recognition to generate student interest, 124 provided classroom instruction, and 75 provided student scholarships or fellowships.

- To support teachers, 115 programs provided curriculum development, 83 programs provided teacher in-service, professional development, or retention activities, and 52 programs provided preservice or recruitment activities.
- To support STEM research, 68 programs reported conducting research to enhance the quality of STEM education.
- To support institutions, 65 programs provided institutional support to management and administrative activities, and 46 programs provided support for expanding the facilities, classrooms, and other physical infrastructure of institutions.

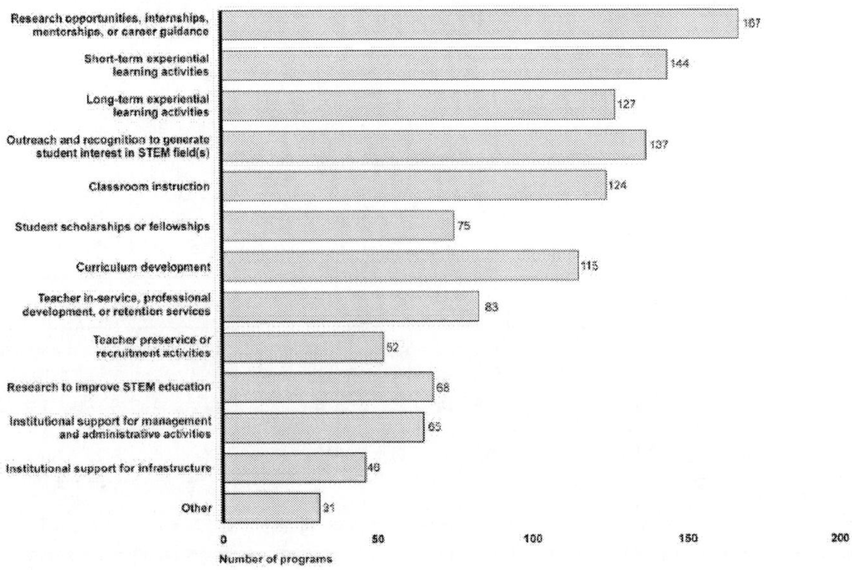

Source: GAO analysis of survey responses.

Figure 5. Services Provided by Federal STEM Education Programs.

Many programs provided similar services to similar target groups. For example, 39 programs that listed chemistry as a primary field of focus provided student scholarships or fellowships to postsecondary students. Many of these programs offered scholarships and fellowships to minority, disadvantaged, or underrepresented students across a broad range of STEM fields. Specifically, some programs, like NASA's Minority University Research and Education Program (MUREP) and the Department of Commerce's Dr. Nancy Foster Scholarship Program, offered scholarships, along with a range of other services, to underrepresented and underserved students in overlapping STEM fields even though the programs focused on preparing students to work in fields that support the science mission of each agency. Overall, most programs provided an array of services to target groups—150 programs provided four or more services, while only 16 programs provide one service.

Similar STEM Fields of Focus

In addition to the serving multiple target groups, most programs also provided services in multiple STEM fields.

Twenty-three programs targeted one specific STEM field, while 121 programs targeted four or more specific STEM fields. In addition, 26 programs indicated not focusing on any specific STEM field, they provided services eligible for use in any STEM field.

Five different STEM fields had over 100 programs that provided services. Biological sciences and technology were the most selected STEM fields focused on by programs. Agricultural sciences, which was the least commonly selected, still had 27 programs that provided services specifically to that STEM field.

Table 2. STEM Fields of Focus and Target Groups of Federal STEM Education Programs

Target groups	Agricultural sciences	Biology	Chemistry	Computer science	Earth sciences	Engineering	Mathematics	Physics	Social sciences	Technology
K-12 students	8	40	36	30	38	32	33	31	19	43
Postsecondary students	22	99	85	84	64	89	79	76	62	87
K-12 teachers	5	36	33	25	39	26	28	29	17	38
College faculty	17	49	42	43	35	47	37	36	30	50

Source: GAO analysis of survey results.

Note: Many STEM education programs serve multiple target groups with multiple STEM fields of focus. The totals cited in table 2 will not sum to 209, the number of programs in our review. Earth sciences includes atmospheric and ocean sciences; social sciences includes psychology, sociology, anthropology, cognitive science, economics, and behavior sciences.

While the data show that many programs had similar target groups and similar STEM fields of focus, it is also important to compare programs' target groups and STEM fields of focus to get a better picture of the potential target beneficiaries that could be served within a given STEM discipline.

For example, both the National Environmental Satellite, Data, and Information Service (NESDIS) Education and the Graduate Automotive Technology Education Program provided scholarships or fellowships to postsecondary students, but one focused on students in earth, atmospheric, and ocean sciences programs, and one on students in engineering, specifically in the areas of hybrid propulsion systems, fuel cells, biofuels, energy storage systems, lightweight materials, and advanced computation; therefore, the target beneficiaries served by these programs are quite different.

Nevertheless, 72 programs provided services to postsecondary students in physics. As table 2 illustrates, many programs offered services to similar target groups in similar STEM fields of focus.

Overlapping programs can lead to individuals and institutions being eligible for similar services in similar STEM fields offered through multiple programs and, without information sharing, could lead to the same service being provided to the same individual or institution.

Similar Objectives

Many STEM education programs had similar objectives. The vast majority (87 percent) of STEM education programs indicated that attracting and preparing students throughout their academic careers in STEM areas was a primary objective.

In addition to attracting and preparing students throughout their academic careers in STEM areas, officials also indicated the following primary program objectives:

- improving teacher education in STEM areas (teacher development)— 26 percent,
- improving or expanding the capacity of K-12 schools or postsecondary institutions to promote or foster education in STEM fields (institution capacity building)—24 percent, and
- conducting research to enhance the quality of STEM education provided to students (STEM education research)—18 percent.

Many programs also reported having multiple primary objectives. While 107 programs focused solely on student education, 82 others indicated having multiple primary objectives, and 9 programs reported having 4 or more primary objectives.

Few programs reported focusing solely on teacher development, institution capacity building, or STEM education research. Most of these objectives were part of a larger program that also focused on attracting and preparing students in STEM education.

Overlapping Programs Are Not Necessarily Duplicative

However, even when programs overlapped, the services they provided and the populations they served may differ in meaningful ways and would therefore not necessarily be duplicative:

- There may be important differences between the specific field(s) of focus and the program's stated goals. For example, both Commerce's National Estuarine Research Reserve System Education Program and the Nuclear Regulatory Commission's Integrated University Program provided scholarships or fellowships to doctoral students in the field of physics. However, the National Estuarine Research Reserve System Education Program's goal was to increase environmental literacy related to estuaries and coastal watersheds by providing students with an opportunity to conduct research of local and national significance that focuses on enhancing coastal zone management; while the Integrated University Program focused on supporting education in nuclear science, engineering, and related fields with the goal of developing a workforce capable of designing, constructing, operating, and regulating nuclear facilities and capable of handling nuclear materials safely.
- Programs may be primarily intended to serve different specific populations within a given target group. For example, 65 programs were primarily intended to serve minority, disadvantaged, or underrepresented groups and 10 programs limited their services to students or teachers in specific geographic areas.[24] Indeed, of the 34 programs providing services to K-12 students in the field of technology, 10 were

primarily intended to serve specific underrepresented, minority, or disadvantaged groups, and 2 were limited geographically to individual cities or universities.

- Furthermore, individuals may receive assistance from different programs at different points throughout their academic careers that provide services that complement or build upon each other, simultaneously supporting a common goal rather than serving cross purposes.

The America COMPETES Reauthorization Act of 2010 Requires Coordination and Strategic Planning of STEM Education Initiatives

Despite past recommendations from ACC and others to improve coordination among STEM education programs, efforts to coordinate STEM education programs across the government remain limited. Although 83 percent of STEM education programs overlapped to some degree with at least one other program, only 33 percent of programs reported coordinating with other agencies that provide similar STEM education services to similar program beneficiaries, not including basic governmentwide inventory efforts. Some program officials mentioned that they coordinate by employing informal mechanisms for information sharing such as conversations and meetings between program staff, sharing resources or best practices, and participating in conferences with other agency officials. Other efforts included developing memorandums of understanding, issuing joint guidance, cofunding programs, and establishing interagency working groups focused on specific science subjects or providing a specific service to a specific target group.

With the growing concern for improved federal coordination and planning in STEM education, Congress passed the America COMPETES Reauthorization Act of 2010, which requires the Director of OSTP to establish a committee under NSTC to coordinate STEM education activities and programs among respective federal agencies and OMB. The NSTC Committee on Science, Technology, Engineering, and Math Education (CoSTEM), comprised of representatives from 11 different federal agencies, convened its first meeting in March 2011. The statute requires NSTC to develop a 5-year governmentwide STEM education strategic plan and identify areas of duplication among federal programs. CoSTEM provides NSTC with an opportunity to improve coordination and be more strategic with the federal investment in STEM education. Best practices in interagency collaboration include developing ongoing mechanisms and processes to monitor, measure, and report agency progress toward NSTC's strategic planning goals and making the results publicly available to improve accountability. According to OSTP officials, a description of the 5-year strategic plan should be publicly available in early 2012; however, as called for in its charter, the committee will terminate no later than March 31, 2015, before the first 5-year plan is carried out, unless it is renewed by the Director of OSTP.

Pursuant to requirements under the 2010 reauthorization of the COMPETES Act, NSTC has implemented several initiatives to enhance coordination. In December 2011, CoSTEM published a report on the inventory of the federal STEM education portfolio that, according to OSTP officials, will be used to improve coordination and inform the strategic planning process.[25] Specifically, OSTP officials said the inventory will allow agencies to identify similar programs and share information and best practices. Without proper coordination, overlapping programs may not share information about the results of the actions taken or research conducted with other interested agencies, possibly leading to numerous programs providing assistance to address the same issue or area of research.

To the extent that CoSTEM identifies duplicative programs, it will be important that it considers the trade-offs associated with program consolidation and assist agencies in determining the most effective and efficient way to reduce duplication. Cost savings might be achieved through the consolidation of duplicative program administrative structures. However, our past work has shown that program consolidation can be more expensive in the short term, and, in the long term, cost savings could be diminished if the workload associated with certain administrative activities remains the same, such as reviewing and assessing applications, providing technical assistance, and monitoring program recipients.[26] Furthermore, over 90 percent of STEM education programs that reported on administrative costs estimated having administrative costs lower than 10 percent of their total program costs.[27] Last, the consolidation of some programs may require congressional action because some programs may be statutorily mandated.

LIMITED USE OF PERFORMANCE MEASURES AND EVALUATIONS MAY HAMPER ABILITY TO ASSESS EFFECTIVENESS

Programs Varied in Their Ability to Provide Information on Populations Served

Program officials varied in their ability to provide reliable information on the number of students, teachers, or institutions directly served by their programs—which is a type of output measure. For example, among programs in our review that served postsecondary teachers and students in 2010, about one-fifth of them did not know the number served. However, depending on the service delivery structure of the program, it may be more difficult to track this number. In some cases, the program's agency did not maintain databases or contracts that would track the number of students served by the program. In other cases, programs may not have been able to provide information on the numbers of institutions they served because they provided grants to secondary recipients. For example, one program indicated that it gives grants to institutions to provide internships or scholarships but that funding goes directly to students, so it does not have information about the number of institutions served. Programs that provide informal educational activities or online services also reported difficulty in tracking the number of individuals who benefited from their programs. The validity and accuracy of the reported output data for some of these programs may be questionable and may hinder program planning and assessment. Programs that reported the numbers they served used varied approaches to collect this information, including annual reports from grant recipients, student enrollment counts, estimates of the expected number of participants reached, and reviews of funding proposals. Some programs had third parties track the numbers served, but did not always take steps to independently verify the data or review the process for how the information was collected. Further, the inconsistent collection of output measures across programs makes it challenging to aggregate the number of students, teachers, and institutions served and to assess the effectiveness of the overall federal effort. Output data are an important component to understanding whether programs are likely to meet their goals.

For example, if a K-12 program has the goal of increasing the number of undergraduates pursuing coursework in STEM fields, it is important to know how many K-12 students were in the program. Without such data, it would be challenging to assess the intended outcome of the program—for example, the number of students who actually went on to pursue such coursework.

Outcome Data Are Not Clearly Reflected in Agencies' Performance Plans and Reports

Agencies in our review did not use outcome measures in a way that is clearly reflected in their performance plans and performance reports— publicly available documents they use for performance planning.[28] This may hinder decisionmakers' ability to assess how agencies' STEM efforts contribute to agencywide performance goals and the overall federal STEM effort. In our review of fiscal year 2010 annual performance plans and reports of the 13 agencies with STEM programs, we found that most agencies did not connect STEM education activities to agency goals or measure and report on the progress of those activities.[29] These documents typically lay out agency performance goals that establish the level of performance to be achieved by program activities during a given fiscal year, the measures developed to track progress, and what progress has been made toward meeting those performance goals.

As figure 6 illustrates, in our review of agencies' specific references to their overall STEM education initiatives, although 38 percent of agencies mentioned STEM education in their performance plans and 62 percent in their performance reports, fewer cited outcome measures related to STEM education. More specifically, in reporting on their progress toward meeting their performance goals, 46 percent of the agencies mentioned STEM education as contributing to one of these goals in their performance reports. Moreover, agencies that spent the most on STEM education were not necessarily more likely to mention, connect to agency performance goals, or measure and report on progress of their STEM efforts. For instance, NASA, which administered 9 STEM programs and obligations of about $209.6 million in fiscal year 2010, mentioned its overall STEM education efforts and connected them to agency performance goals in its planning documents and measured and reported on progress in both its performance plan and report. On the other hand, HHS's National Institutes of Health, which administered the most STEM education programs (44) and obligations of about $573.6 million, referred to agency performance goals and outcome measures of its STEM education efforts only in some of its institutes' performance reports, but not in its NIH-wide performance plan.

As figure 7 illustrates, in our review of agencies' specific references to their STEM education programs, while the 13 agencies combined mentioned 38 percent of their programs in their performance plans, they connected 19 percent of their STEM education programs to agency performance goals and measured and reported on progress of 9 percent of the programs. Agencies' STEM education obligations and number of programs did not correlate directly with their likelihood of connecting the programs to agency performance goals or measuring and reporting on their progress in performance plans and reports.

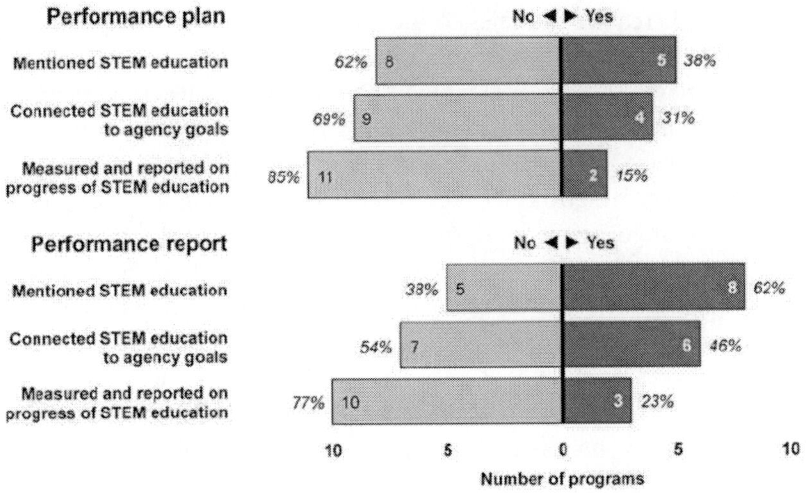

Source: GAO analysis of GAO survey data.

Figure 6. Integration of Overall STEM Education Efforts in Agencies' Performance Plans and Reports.

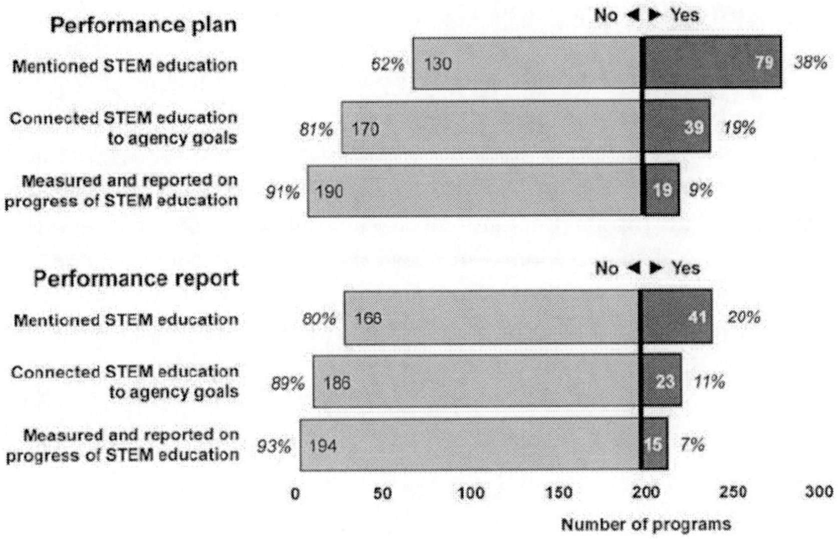

Source: GAO analysis of GAO survey data.

Figure 7. Integration of STEM Education Programs in Agencies' Performance Plans and Reports.

For example, Interior, through the U.S. Geological Survey, which administered just 3 STEM education programs in fiscal year 2010, mentioned all of its programs in its performance plan. In contrast, NSF, which administered 37 STEM education programs and obligated about $1.1 billion in fiscal year 2010, connected only 2 of its programs to agency performance goals while measuring and reporting on progress in its performance plan and report.

The GPRA Modernization Act of 2010 and the America COMPETES Reauthorization Act of 2010 afford agencies the opportunity to better utilize performance measures for both governmentwide and agency-specific STEM education efforts. For example, the GPRA

Modernization Act will require agencies to identify program activities and other activities, which may include STEM education activities that contribute to each performance goal. It recognizes the importance of governmentwide performance goals as it requires OMB to develop, in coordination with agencies, long-term, crosscutting federal government priority goals that are to be updated or revised every 4 years, which will be tracked quarterly in order to review progress to improve government performance. According to OMB guidance, it will announce interim federal government priority goals in February 2012 and finalize its goals in February 2014. The America COMPETES Reauthorization Act of 2010 also focuses on accountability through strategic planning, and has specific requirements for agencies with STEM programs. Specifically, it requires NSTC to develop a STEM education strategic plan with long-term objectives, metrics to assess agencies' progress, and approaches taken by participating agencies to assess the effectiveness of their STEM programs and activities. However, while OSTP will be required to report on agencies' annual progress toward the long-term objectives, an OSTP official said there is no mechanism to make agencies align their performance measures with the goals and objectives in the strategic plan.

Most STEM Education Programs Have Not Completed Evaluations

Little is known about the effectiveness and performance of STEM education programs because the majority of them (66 percent) have not conducted an evaluation of their entire program since 2005 (as figure 8 illustrates). We define "evaluation" as an individual systematic study conducted periodically or on an ad hoc basis to assess how well a program is working, typically relative to its program objectives. Some programs that reported that they did not complete an evaluation reported they had their grantees complete one; however, in those cases, few programs used these grantee evaluations to inform a more comprehensive evaluation of the entire program that they or an external evaluator completed.

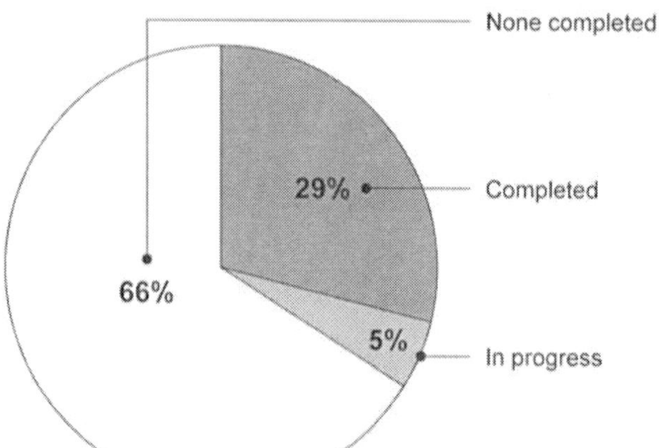

Source: GAO analysis of survey responses.

Note: Of programs in our review, 12 percent (26 of 209) reported they planned to complete one, and 12 percent (26 of 209) used other reporting and evaluation activities such as memos summarizing program activities and evaluation planning documents.

Figure 8. Percentage of STEM Education Programs, by Status of Evaluations since 2005.

In total, since 2005, agencies conducting 61 programs, (representing about 61 percent of the $3.1 billion obligated in fiscal year 2010) responded that they had completed evaluations—all of which used a variety of methods and designs. We reviewed evaluations for 35 of the 61 programs.[30] Most of the 35 program evaluations we reviewed used methods and designs that appropriately assessed how well they met their stated objectives. For instance, one evaluation selected a random sample of its former program participants and compared them with a sample of students who had applied to the same program, but had not participated. While former participants had some statistically significant academic outcomes when compared with the nonparticipants, the evaluation also noted other factors that may have influenced the favorable outcomes of the program—for example, that participants, on average, were more interested in careers in science and math than the nonparticipants, so the true effects of program participation may be overstated.

Even though most of the 35 programs we reviewed employed appropriate methods and designs to assess their programs' effectiveness, we identified several ways to improve evaluations of STEM education, based on our review.

- Improved survey response rates: Many of the evaluations we reviewed had low response rates. Without better response rates, generalizations from the results may be limited.
- Better alignment of the methods with other components of the evaluation: Specifically, 10 of the programs used evaluation methods that were not fully aligned with the evaluation questions and the program context.[31] For example, 3 of these evaluations had data limitations, thus hindering the use of methods that could collect the full range of data to inform program outcomes.
- Robust use of criteria to measure outcomes: Among the 27 programs that measured outcomes, 9 did not evaluate them against any criteria. Without criteria to evaluate the outcomes, it may be difficult to establish programmatic impact and assess performance and effectiveness.

Furthermore, in order to influence program practice, the evaluation results must be disseminated widely. While nearly all of the STEM education programs that reported completing an evaluation reported using different mechanisms to disseminate results, they did not always share results in a way that facilitated knowledge sharing. Program officials reported that the most common means of dissemination of their results were through their websites or at conferences or forums, which, according to a 2006 NSTC report, were methods that require practitioners to actively seek out results, so such methods may prevent the results of the research from being conveyed to them. However, these mechanisms have limits. For example, NSTC also reported that STEM education research results may not be conveyed to practitioners because the results often lack applicability, some are ambiguous, and the culture of teaching typically does not make decisions based on research findings. NSTC identified other issues with sharing information about STEM education program results and suggested several actions that agencies could take to improve dissemination, such as engaging practitioners to collaborate with researchers in setting research agendas.[32] According to NSTC officials, most agencies do not share or disseminate evaluations in a way that could be useful for coordination.

Conclusion

Although the federal government invests billions of dollars annually in STEM education programs, there remains concerns over U.S. economic and educational competitiveness, particularly with regard to the national educational system's ability to produce citizens literate in STEM subjects and to produce future scientists, technologists, engineers, and mathematicians. Prior reports on STEM education highlighted the lack of federal governmentwide planning and coordination. Recently, both Congress and the administration called for a more strategic and effective approach to the federal government's investment in STEM education. The America COMPETES Reauthorization Act of 2010 requires the Director of OSTP to establish a committee under NSTC to develop a 5-year strategic plan and submit annual reports, including a description of the plan, to Congress. The plan is expected to include common measures to assess progress toward the plan's goals. In addition, the GPRA Modernization Act of 2010 requires agencies to identify program activities that contribute to each performance goal, and, as agencies implement this provision, more information about STEM education efforts in performance plans and reports can be expected. NSTC's ongoing strategic planning efforts provide an opportunity to develop guidance on how to incorporate STEM- and program-specific education goals and measures in agencies' performance planning and reporting process and align their STEM education efforts with a governmentwide STEM education strategy. To further strengthen strategic planning and coordination efforts, an accountability and reporting framework should exist to ensure agencies are adhering to NSTC's strategic plan.

While the STEM education programs we reviewed in this report are fragmented and overlapping to some degree, they are not necessarily duplicative of one another. More analysis is needed to identify areas of duplication among federal STEM education programs and ensure that the federal investment in these programs advances NSTC's 5-year strategic plan that is under development. In this era of budget constraints, governmentwide strategic planning can play a critical role in addressing concerns about program fragmentation, overlap and duplication.

Fragmentation and overlap can (1) frustrate federal officials' efforts to administer programs in a comprehensive manner, (2) limit the ability to determine which programs are most cost-effective, and (3) ultimately increase program administrative costs. Therefore, if NSTC's 5-year strategic plan is not developed in a way that aligns agencies' efforts to achieve governmentwide goals, enhances the federal government's ability to assess what works, and concentrates resources on those programs that advance the strategy, the federal government may spend limited funds in an inefficient and ineffective manner that does not best help to improve the nation's global competitiveness.

Understanding program performance and effectiveness is also key in determining where to strategically invest limited federal funds to achieve the greatest impact in developing a pipeline of future workers in STEM fields. Programs need to be appropriately evaluated to determine what is working and how improvements can be made.

However, most agencies have not conducted comprehensive evaluations since 2005 to assess the effectiveness of their STEM education programs. Furthermore, methods for dissemination of program evaluations, especially to practitioners, could be improved. Agency and program officials would benefit from guidance and information sharing within and across agencies about what is working and how to best evaluate programs. This could not only help to improve individual program performance, but also inform agency and governmentwide decisions about which programs should continue to be funded.

Without an understanding of what is working in some programs, it will be difficult to develop a clear strategy for how to spend limited federal funds.

RECOMMENDATIONS FOR EXECUTIVE ACTION

The Director of OSTP should direct NSTC to

1) Develop guidance for how agencies can better incorporate each agency's STEM education efforts and the goals from NSTC's 5-year STEM education strategic plan into each agency's own performance plans and reports.
2) Develop a framework for how agencies will be monitored to ensure that they are collecting and reporting on NSTC strategic plan goals. This framework should include alternatives for a sustained focus on monitoring coordination of STEM programs if the NSTC Committee on STEM terminates in 2015 as called for in its charter.
3) Work with agencies, through its strategic planning process, to identify programs that might be candidates for consolidation or elimination. Specifically, this could be achieved through an analysis that includes information on program overlap, similar to the analysis conducted by GAO in this report, and information on program effectiveness. As part of this effort, OSTP should work with agency officials to identify and report any changes in statutory authority necessary to execute each specific program consolidation identified by NSTC's strategic plan.
4) Develop guidance to help agencies determine the types of evaluations that may be feasible and appropriate for different types of STEM education programs and develop a mechanism for sharing this information across agencies. This could include guidance and sharing of information that outlines practices for evaluating similar types of programs.

AGENCY COMMENTS AND OUR EVALUATION

We provided a draft of this report to the Office of Science and Technology Policy (OSTP) and the Office of Management and Budget (OMB) for review and comment. OSTP provided technical comments that we incorporated as appropriate. OMB had no concerns with the report.

George A. Scott
Director, Education, Workforce,
and Income Security Issues

APPENDIX I: OBJECTIVES, SCOPE, AND METHODOLOGY

The objectives of our report were to determine (1) the number of federal agencies and programs that provided funding for science, technology, engineering, and mathematics (STEM) education programs in fiscal year 2010; (2) the extent to which STEM programs have similar objectives, serve similar target groups, provide similar types of services, and, if necessary, what opportunities exist to increase coordination; and (3) the extent to which STEM programs have measured their effectiveness. To inform all of our objectives, we reviewed relevant federal laws and regulations. We also reviewed previous work that was conducted to catalog and assess the federal investment in STEM education programs, including a 2005 GAO study,[1] the 2007 Academic Competitiveness Council (ACC) report,[2] and the 2010 Office of Management and Budget (OMB) inventory. We reviewed relevant literature and past reports on STEM education, including the 2010 President's Council of Advisors on Science and Technology (PCAST) report entitled *Report to the President: Prepare and Inspire: K-12 Education in Science, Technology, Engineering, and Math (STEM) for America's Future* [3] and the National Academies Press report entitled *Rising above the Gathering Storm: Energizing and Employing America for a Brighter Economic Future: Committee on Prospering in the Global Economy of the 21st Century: An Agenda for American Science and Technology.*[4]

In addition, we interviewed officials from OMB, the Office of Science and Technology Policy (OSTP), and 13 other federal agencies that administer STEM education programs to gather information on their STEM education efforts, the extent of coordination between programs, and the existence of program evaluations. We attended several STEM education conferences to gather additional perspectives about federal STEM education programs. Finally, we reviewed evaluations provided by program officials as well as agencies' annual performance plans and reports.

To gather information on federal STEM education programs and to assess the level of fragmentation, overlap, and potential duplication among them, we first reviewed past GAO work on assessing the level of fragmentation, overlap, and duplication among other groups of federal programs. Next, we surveyed over 200 programs across 13 agencies that met our definition of a STEM education program (see below) with questions about program objectives, target populations, services provided, interagency coordination, outcome measures and evaluations, and funding information.

In December 2011, NSTC's Committee on STEM Education released its inventory of the federal STEM education portfolio.[5] The NSTC inventory differs from GAO's survey in that it counts investments and allocations, whereas GAO asked agencies to report on programs and obligations.

Definition of STEM Education Program

For the purposes of our study, we defined a federally funded STEM education program as a program funded in fiscal year 2010 by congressional appropriation or allocation that includes one or more of the following as a primary objective:

- attract or prepare students to pursue classes or coursework in STEM areas through formal or informal education activities (informal education programs provide support for activities provided by a variety of organizations that offer students learning opportunities outside of formal schooling through contests, science fairs, summer programs, and other means; outreach programs targeted to the general public should not be included),
- attract students to pursue degrees (2-year, 4-year, graduate, or doctoral degrees) in STEM fields through formal or informal education activities,
- provide training opportunities for undergraduate or graduate students in STEM fields (this can include grants, fellowships, internships, and traineeships that are targeted to students; general research grants that are targeted to researchers that may hire a student to work in the lab should not be considered a STEM education program),
- attract graduates to pursue careers in STEM fields,
- improve teacher (preservice or in-service) education in STEM areas,
- improve or expand the capacity of K-12 schools or postsecondary institutions to promote or foster education in STEM fields, or
- conduct research to enhance the quality of STEM education programs provided to students.[6]

In addition, we defined STEM education programs to include grants, fellowships, internships, and traineeships. While programs designed to retain current employees in STEM fields were not included, programs that fund retraining of workers to pursue a degree in a STEM field were included because these programs help increase the number of students and professionals in STEM fields by helping retrain non-STEM workers to work in STEM fields.

For the purposes of this study, we defined the term "program" as an organized set of activities supported by a congressional appropriation or allocation. Further, we defined a program as a single program even when its funds were allocated to other programs as well. We asked agency officials to provide a list of programs that received funds in fiscal year 2010. This included programs that received one-time, limited funds in fiscal year 2010, such as earmarks.[7]

Definition of STEM Fields

We determined that a STEM field should be considered any of the following broad disciplines:

- agricultural sciences;
- biological sciences;
- chemistry;
- computer science;
- earth, atmospheric, and ocean sciences;
- engineering;
- mathematics;
- physics;

- social sciences (e.g., psychology, sociology, anthropology, cognitive science, economics, behavioral sciences); or
- technology.

In addition, we determined that our definition of STEM education would include health care programs that train students for careers that are primarily in scientific research. We did not, however, include health care programs that train students for careers that are primarily in patient care, that is, those that trained nurses, doctors, dentists, psychologists, or veterinarians.

Program Selection

To identify federally funded STEM education programs, first we developed a combined list of programs based on the findings of three previous STEM education inventory efforts completed by GAO in 2005, ACC in 2007, and OMB in 2010. Second, we shared our list with agency officials, provided our definition of STEM education program, and asked officials to make an initial determination about which programs should remain on the list and which programs should be added to the list. If agency officials indicated they wanted to remove a program from our list, we asked for additional information. For example, programs on our initial list may have been terminated or consolidated, or did not receive federal funds in fiscal year 2010. In addition, we asked officials to provide program descriptions, program names, and contact information. Next, we reviewed each agency's submission and individual program information and determination. We also gathered additional information on the program, mainly through agency websites and program materials, and held discussions with program officials to understand the program in more detail.

Table 3. Reasons for Exclusion of STEM Education Activities from Our Survey

Exclusion rationale	Number of programs
Program did not receive congressional appropriation or allocation in fiscal year 2010.	59
Program was consolidated or part of another program.	48
Program focused on professionals or postdoctoral positions.	41
Entry was duplicative or not recognized by agency officials.	34
Nonprogrammatic STEM activities.	24
Program for which STEM is not a primary purpose.	17
Research-related program, not focused on STEM education.	8
Program is focused on patient care.	8
General awareness program not focused on students or teachers.	1
Total entries excluded	240

Source: GAO.

On the basis of this additional information, we excluded programs that we found did not meet our definition of a STEM education program. Once our determinations were made, we asked each agency to confirm the list of programs and the names and contact information for the officials who would be responsible for completing the survey. In total, we determined that 274 programs should receive a survey.[8] We also included several screening questions in the survey to provide an additional verification to ensure the programs met our definition of a STEM education program. Nineteen programs did not pass our screening questions and therefore were excluded from our analysis. All in all, 209 programs were included in our final analysis. For a list of the 209 STEM education programs by agency, see appendix II. For a summary of excluded programs and their exclusion rationales, see table 3. Furthermore, we provide aggregate survey responses from these programs in an e-supplement (GAO-12-110SP).

Survey

Design and Implementation

We developed a web-based survey to collect information on federal STEM education programs. See GAO-12-110SP for a copy of the survey's full text. The survey included questions on program objectives, target groups served, services provided, academic fields of focus, output metrics, outcome measures, obligations, and program evaluations. To minimize errors arising from differences in how questions might be interpreted and to reduce variability in responses that should be qualitatively the same, we conducted pretests with 14 different programs in March and April 2011. To ensure that we obtained a variety of perspectives on our survey, we selected 14 programs from 11 different agencies that differed in program scope, objectives, services provided, target groups served, evaluations completed, and funding sources. We included budget staff as well as program officials in the pretests to ensure budget-related terms in the survey were understandable and available. An independent GAO reviewer also reviewed a draft of the survey prior to its administration. On the basis of feedback from these pretests and independent review, we revised the survey in order to improve its clarity.

After completing the pretests, we administered the survey. On May 3, 2011, we sent an e-mail announcement of the survey to the officials responsible for the programs selected for our review, notifying them that our online survey would be activated within a week. On May 11, 2011, we sent a second e-mail message to officials that informed them that the survey was available online. In that e-mail message, we also provided them with unique passwords and usernames. We made telephone calls to officials and sent them follow-up e-mail messages, as necessary, to clarify their responses or obtain additional information. We received completed surveys from 269 programs, for a 100 percent response rate. We collected survey responses through August 31, 2011.

Analysis of Responses and Data Quality

We used standard descriptive statistics to analyze responses to the survey. Because this was not a sample survey, there were no sampling errors. To minimize other types of errors, commonly referred to as nonsampling errors, and to enhance data quality, we employed survey design practices in the development of the survey and in the collection, processing, and analysis of the survey data. For instance, as previously mentioned, we pretested the survey with federal officials to minimize errors arising from differences in how questions might be interpreted and to reduce variability in responses that should be qualitatively the same. We further reviewed the survey to ensure the ordering of survey sections was appropriate and that the questions within each section were clearly stated and easy to comprehend. To reduce nonresponse bias, another source of nonsampling error, we sent out e-mail reminder messages to encourage officials to complete the survey. In reviewing the survey data, we performed automated checks to identify inappropriate answers. We further reviewed the data for missing or ambiguous responses and followed up with agency officials when necessary to clarify their responses. To assess output measures, we asked a series of questions to assess the agency's procedures, policies, and internal controls to ensure the quality of data provided in the survey.

For program obligations questions, we sampled 10 percent of responses reviewing documentary evidence to corroborate survey responses. For evaluation questions, we reviewed program evaluations provided to corroborate survey responses. To assess the reliability of data provided in our survey, we incorporated questions about the reliability of the programs' data systems, reviewed documentation for a sample of selected questions, conducted internal reliability checks, and conducted follow-up as necessary.

While we did not verify all responses, on the basis of our application of recognized survey design practices and follow-up procedures, we determined that the data used in this report were of sufficient quality for our purposes. We did not report on data that we found of questionable reliability based on our review of data reliability questions in the survey—such as the number of students and teachers served. All data analysis programs were also independently verified by a GAO data analyst for accuracy.

Performance Evaluations

Program officials who responded on their survey that an evaluation of their program had been completed in 2005 or later provided us with information about their most recent evaluations. GAO defines "evaluation" as an individual systematic study conducted periodically or on an ad hoc basis to assess how well a program is working. Studies are often conducted by experts external to the program, inside or outside the agency, as well as by program managers.

Furthermore, an evaluation typically examines achievement of program objectives in the context of other aspects of program performance or in the context in which it occurs.[9] After ensuring that the evaluations met this definition, we reviewed them to analyze their characteristics, including their methods and designs, and the extent to which program outcomes were measured. In addition, we examined whether the methods and designs were appropriate given the evaluation questions and program context.

In total, 61 programs responded that they had completed a program evaluation since 2005, and we reviewed evaluations from 35 of those programs. Because we requested that officials provide us with a citation for the most recent evaluation, we selected the most recent one for our review. We did not review evaluations from the remaining 26 programs for a variety of reasons.

Specifically, they were committee of visitors reports, and other types of reports that did not have evaluation information that aligned with the criteria by which we analyzed the other evaluations. Among these reports, we were unable to obtain 6 of them. As a result, we were unable to analyze them and determine whether they met GAO's definition of evaluation.

For more details about the evaluations in our review, see appendix III.

Agencies' Annual Performance Plans and Reports

We reviewed agencies' fiscal year 2010 required strategic planning documents—performance plans and performance reports—to determine the extent to which they incorporated program-specific and broad-based STEM goals and objectives.[10] The performance plans and reports were done at the agency level, while others were done at other levels, such as the institute or office level—in which case we reviewed the documents that covered the particular STEM program(s) in our review. When reviewing these documents, we determined the extent to which

- agencies made any reference to agencywide STEM initiatives or particular STEM education programs, in general, but not in the context of agency goals or of outcome measures;
- agencies connected their STEM initiatives or their individual STEM programs to agency goals; and
- agencies articulated outcome measures of their STEM initiatives or of individual STEM programs.

APPENDIX II: LIST OF STEM EDUCATION PROGRAMS WITH FISCAL YEAR 2010 OBLIGATIONS

Agency	Program	Fiscal Year 2010 STEM education program obligations[a]
NASA	Aeronautics Research Directorate-STEM Education activities	$4,153,000
	Exploration Systems Directorate-STEM Education activities	6,400,000
	Higher Education	18,346,329
	K-12 STEM Program	36,291,069
	Minority University Research and Education Program	28,862,619
	NASA Informal Education Opportunities (NIEO)	14,295,934
	NASA Science Mission Directorate E/PO	30,057,100
	Space Grant/EPSCoR Program	68,910,696
	Space Operations Directorate-STEM Education activities	2,293,000
National Science Foundation	Advanced Technological Education (ATE)	64,510,000
	Alliances for Graduate Education and the Professoriate (AGEP)	16,730,000
	Broadening Participation in Computing (BPC)	14,000,000
	Centers for Ocean Science Education Excellence	5,700,000
	CISE Pathways to Revitalized Undergraduate Computing Education (CPATH)	4,370,000
	Cyberinfrastructure Training, Education, Advancement, and Mentoring for Our 21st Century Workforce (CI-TEAM)	4,850,000
	Discovery Research K-12 (DR-K12)	118,380,000
	East Asia & Pacific Summer Institutes for U.S. Graduate Students (EAPSI)	1,740,000
	Engineering Education (EE)	13,740,000
	Enhancing the Mathematical Sciences Workforce in the 21st Century (EMSW21)	15,070,000
	Ethics Education in Science & Engineering (EESE)	2,650,000
	Federal Cyber Service: Scholarship for Service (SFS)	14,870,000
	Geoscience Education	2,020,000

(Continued)

Agency	Program	Fiscal Year 2010 STEM education program obligations[a]
	Geoscience Teacher Training (GEO-Teach)	2,980,000
	Global Learning and Observations to Benefit the Environment (GLOBE)	1,100,000
	Graduate Research Fellowship Program (GRFP)	136,130,000
	Graduate STEM Fellows in K-12 Education Program (GK-12)	55,970,000
	Historically Black Colleges and Universities Undergraduate Program (HBCU-UP)	32,060,000
	Informal Science Education (ISE)	65,850,000
	Integrative Graduate Education and Research Traineeship (IGERT) Program	69,700,000
	Interdisciplinary Training for Undergraduates in Biological and Mathematical Sciences (UBM)	2,700,000
	International Research Experiences for Students (IRES)	3,430,000
	Louis Stokes Alliances for Minority Participation (LSAMP)	44,550,000
	Math and Science Partnership	57,930,000
	Nanotechnology Undergraduate Education in Engineering	1,830,000
	Opportunities for Enhancing Diversity in the Geosciences	4,180,000
	Polar Education Program	1,500,000
	Research and Evaluation on Education in Science and Engineering (REESE)	45,670,000
	Research Experiences for Teachers (RET) in Engineering and Computer Science	5,410,000
	Research Experiences for Undergraduates (REU)	80,990,000
	Research in Disabilities Education (RDE)	6,920,000
	Research on Gender in Science and Engineering (GSE)	11,570,000
	Robert Noyce Teacher Scholarship Program	54,930,000
	Science, Technology, Engineering, and Mathematics Talent Expansion Program (STEP)	31,640,000
	Transforming Undergrad Education in STEM (TUES)	41,600,000
	Tribal Colleges and Universities Program (TCUP)	13,350,000
	Undergraduate Research and Mentoring in the Biological Sciences (URM)	9,000,000
Nuclear Regulatory Commission	Integrated University Program	15,000,000
	Minority Serving Institutions Program (MSIP)	2,838,500
	Nuclear Education Curriculum Development Grants	4,700,997
Department of Agriculture		
Animal and Plant Health Inspection Service (APHIS)	AgDiscovery Program	15,000
National Institute of Food and Agriculture (NIFA)	1890 Institution Teaching, Research and Extension Capacity Building Grants Program	17,167,994
	Agriculture in the Classroom	314,912
	Food and Agricultural Sciences National Needs Graduate and Postdoctoral Fellowships Grants Program	3,664,127
	Higher Education Challenge Grants Program	5,654,000
	Higher Education Multicultural Scholars Program	1,126,000
	Hispanic Education Partnership Grants	8,809,568
	New Era Rural Technology Competitive Grants Program	875,000
	Resident Instruction Grants for Institutions of Higher Education in Insular Areas	859,547
	Secondary Education, Two-Year Postsecondary Education and Agriculture in the K-12 Classroom Grants	983,000

Agency	Program	Fiscal Year 2010 STEM education program obligations[a]
Office of the Assistant Secretary for Departmental Management	USDA/1890 National Scholars Program	2,398,947
Department of Commerce		
National Institute of Standards and Technology(NIST)	NIST Summer Institute for Middle School Science Teachers	300,000
	Recovery Act Measurement Science and Engineering Fellowship Program	20,000,000
	Summer Undergraduate Research Fellowship (SURF) Program	595,641
National Oceanic and Atmospheric Administration (NOAA)	Bay Watershed Education and Training (B-WET) Program	9,700,000
	Climate Communications and Education Program	536,000
	Coral Reef Conservation Program	838,000
	Dr. Nancy Foster Scholarship Program	603,125
	Educational Partnership Program with Minority Serving Institutions	14,309,000
	Environmental Literacy Grants	10,388,185
	Ernest F. Hollings Undergraduate Scholarship Program	6,450,638
	Global Learning and Observations to Benefit the Environment	3,000,000
	National Environmental Satellite, Data, and Information Service (NESDIS) Education	2,700,000
	National Estuarine Research Reserve System Education Program	1,020,000
	National Marine Fisheries Service (NMFS) Education	3,084,750
	National Marine Sanctuaries Education Program	908,150
	National Ocean Service (NOS) Education	426,000
	National Sea Grant College Program-Education Component	9,378,529
	National Weather Service Outreach Program	3,070,000
	Teacher at Sea Program	600,000
Department of Defense		
Air Force	Awards to Stimulate and Support Undergraduate Research Experience (ASSURE)	4,500,000
	National Defense Science and Engineering Graduate (NDSEG) Fellowship	38,695,132
	University NanoSatellite Program	660,000
Army	Army Educational Outreach Program (AEOP)	7,885,000
	Consortium Research Fellows Program (CRFP)	1,634,050
	National Science Center (NSC)	1,982,000
Office of the Secretary ofDefense	Autonomous Robotic Manipulation (ARM)	8,180,000
	Computer Science in Science, Technology, Engineering, and Mathematics Education (CS-STEM)	2,661,000
	DoD STARBASE Program	20,000,000
	ENGAGE	2,100,000
	National Defense Education Program (NDEP) K-12	13,595,000
	National Defense Education Program (NDEP) Science, Mathematics And Research for Transformation (SMART)	47,400,000
Military Health System	Uniformed Services University of the Health Sciences (USUHS)	447,000
Navy	Historically Black College and Universities/Minority Institutions Research Education Partnership	700,000
	Iridescent Learning	810,000
	Science and Engineering Apprentice Program (SEAP)	755,000
	SeaPerch	700,000

(Continued)

Agency	Program	Fiscal Year 2010 STEM education program obligations[a]
	The Naval Research Enterprise Intern Program (NREIP)	1,960,000
	University / Laboratory Initiative (ULI)	2,350,000
Department of Education	Developing Hispanic-Serving Institutions: STEM and Articulation Programs (mandatory)	0b
	Graduate Assistance in Areas of National Need	31,005,248
	Mathematics and Science Partnerships	180,478,000
	Minority Science and Engineering Improvement Program	9,503,000
	National Science and Mathematics Access to Retain Talent Program	379,775,972
	Predominantly Black Institutions Competitive Grant Program	0b
	Research in Special Education	11,000,000
	Research, Development, and Dissemination	39,986,940
	Teachers for a Competitive Tomorrow: Baccalaureate Degrees in STEM and Critical Foreign Languages	1,092,000
	Teachers for a Competitive Tomorrow: Master's Degrees in STEM and Critical Foreign Languages	1,092,000
	Upward Bound Math-Science	34,873,057
	Women's Educational Equity	2,423,000
Department of Energy	Academies Creating Teacher Scientists (DOE Acts)	3,721,600
	Advanced Vehicle Competitions	*2,000,000*
	American Chemical Society Summer School in Nuclear and Radiochemistry	546,813
	ASCR-ORNL Research Alliance in Math and Science	250,000
	Community College Institute of Science and Technology	685,000
	Computational Science Graduate Fellowship	7,800,000
	Faculty and Student Teams	1,019,000
	Fusion Energy Sciences Graduate Fellowship Program	800,000
	Graduate Automotive Technology Education	1,000,000
	Hampton University Graduate Studies	48,000
	HBCU Mathematics, Science & Technology, Engineering and Research Workforce Development Program	8,967,507
	Industrial Assessment Centers	6,086,000
	Integrated University Program	5,000,000
	Laboratory Equipment Donation Program	150,000
	Mickey Leland Energy Fellowship	700,000
	Minority Serving Institutions Program	840,000
	Minority University Research Associates Program (MURA)	591,880
	National Science Bowl	2,449,900
	National Undergraduate Fellowship Program in Plasma Physics and Fusion Energy Sciences	370,000
	Office of Science Graduate Fellowship (SCGF) program	17,500,000
	Pan American Advanced Studies Institute	200,000
	Plasma/Fusion Science Educator Programs	779,000
	Pre-Service Teacher Program	429,000
	QuarkNet	750,000
	Science Undergraduate Laboratory Internships	3,802,500
	Solar Decathlon	5,000,000
	Summer Applied Geophysical Experience (SAGE)	100,000
	Technical Career Intern Program	0c
	Wind for Schools	630,000

Agency	Program	Fiscal Year 2010 STEM education program obligations[a]
Department of Health andHuman Services		
Health Resources and Services Administration	Health Careers Opportunity Program	22,086,000
	Public Health Traineeship Program	1,510,000
National Institutes of Health	Bridges to the Baccalaureate Program	6,460,988
	Bridges to the Doctorate	2,977,075
	Cancer Education Grants Program	6,756,869
	Cancer Research Interns	191,608
	CCR/JHU Master of Science in Biotechnology Concentration in Molecular Targets and Drug Discovery Technologies	445,000
	Clinical Research Training Program	1,100,000
	Community College Summer Enrichment Program	105,000
	Curriculum Supplement Series	341,849
	Education Programs for Population Research (R25)	750,154
	Graduate Program Partnerships	16,720,000
	Initiative for Maximizing Student Development	21,412,146
	Intramural NIAID Research Opportunities	129,111
	MARC U-STAR NRSA Program	20,386,651
	Material Development for Environmental Health Curriculum	1,544,868
	National Cancer Institute Cancer Education and Career Development Program	20,442,233
	NCRR Science Education Partnership Award (SEPA)	16,653,015
	NHLBI Minority Undergraduate Biomedical Education Program	475,970
	NIAID Science Education Awards	1,069,978
	NIDDK Education Program Grants	432,000
	NIH Academy	249,866
	NIH Summer Research Experience Programs	1,679,422
	NINDS Diversity Research Education Grants in Neuroscience	821,800
	NLM Institutional Grants for Research Training in Biomedical Informatics	10,143,676
	Office of Science Education K-12 Program	2,270,151
	Post-baccalaureate Intramural Research Training Award Program	24,810,000
	Postbaccalaureate Research Education Program (PREP)	5,780,503
	Recovery Act Limited Competition: NIH Challenge Grants in Health and Science Research	4,953,293
	Research Scientist Award for Minority Institutions	82,146
	Research Supplements to Promote Diversity in Health-Related Research	68,981,252
	RISE (Research Initiative for Scientific Enhancement)	24,441,722
	Ruth L. Kirschstein National Research Service Award Institutional Research Training Grants** (T32, T35)	230,840,328
	Ruth L. Kirschstein NRSA for Individual Predoctoral Fellows, including Underrepresented Racial/Ethnic Groups, Students from Disadvantaged Backgrounds	56,882,642
	Science Education Drug Abuse Partnership Award	2,294,996
	Short Courses in Integrative and Organ Systems Pharmacology	665,937
	Short Courses on Mathematical, Statistical, and Computational Tools for Studying Biological Systems	695,460

(Continued)

Agency	Program	Fiscal Year 2010 STEM education program obligations[a]
	Short Term Educational Experiences for Research (STEER) in the Environmental Health Sciences for Undergraduates and High School Students	568,298
	Short-Term Research Education Program to Increase Diversity in Health-Related Research	4,188,763
	Student Intramural Research Training Award Program	5,868,500
	Summer Genetics Institute	53,935
	Summer Institute for Training in Biostatistics	1,449,092
	Technical Intramural Research Training Award	2,240,000
	Training in Computational Neuroscience: From Biology to Model and Back Again	1,443,450
	Training in Neuroimaging: Integrating First Principles and Applications	1,356,252
	Undergraduate Scholarship Program for Individuals from Disadvantaged Backgrounds	2,426,137
Department of HomelandSecurity		
Science and Technology Directorate	HS-STEM Career Development Grants Program	2,300,000
	HS-STEM Scholars Program	1,920,000
	HS-STEM Summer Internship Program	363,000
	Minority Serving Institutions-Scientific Leadership Awards	2,400,000
	Minority Serving Institutions-Summer Research Team	116,000
Department of the Interior		
U.S. Geological Survey(USGS)	EDMAP Component of the National Cooperative Geologic Mapping Program	566,161
	National Association of Geoscience Teachers (NAGT)-USGS Cooperative Summer Field Training Program	200,000
	Student Intern in Support of Native American Relations (SISNAR)	204,013
Department of Transportation		
Federal Aviation Administration (FAA)	Joint University Program	300,000
	National Center of Excellence for Aviation Operations Research (NEXTOR)	5,393,000
Federal Highway Administration (FHWA)	Garrett A. Morgan Technology and Transportation Education Program	1,250,000
	National Summer Transportation Institute Program	2,602,999
	Summer Transportation Internship Program for Diverse Groups	1,425,000
Research and Innovative Technology Administration(RITA)	University Transportation Centers Program	83,370,600
Environmental Protection Agency	Cooperative Training in Environmental Sciences Research	1,593,184
	Environmental Education Grants	3,450,882
	EPA Greater Research Opportunities (GRO) Fellowships for Undergraduate Environmental Study	1,532,099
	EPA Marshall Scholars Program	205,888
	National Environmental Education and Training Partnership	2,259,500
	National Network for Environmental Management Studies Fellowship Program	469,403

Agency	Program	Fiscal Year 2010 STEM education program obligations[a]
	P3 Award: National Student Design Competition for Sustainability	2,000,000
	President's Environmental Youth Awards	50,000
	Science to Achieve Results Graduate Fellowship Program	6,387,830
	University of Cincinnati/EPA Research Training Grant	$333,153

Source: GAO analysis of survey results.

[a] This number equals the total program obligations for fiscal year 2010, unless the survey respondent provided obligations for the STEM only activities within the program.

[b] Program funding was authorized in 2010, but was not obligated until 2011.

[c] Fiscal year 2010 obligations for the Technical Career Intern Program are reflected in the Mickey Leland Program.

APPENDIX III: REVIEW OF EVALUATIONS

Different types of evaluation designs can provide rigorous evidence of effectiveness if designed well and implemented with a thorough understanding of their vulnerability to potential sources of bias. There are four main types of evaluations that GAO has identified:

- Implementation evaluations (which assess the extent to which the program is operating as intended),
- Impact evaluations (which include experimental and quasi-experimental designs),
- Outcome evaluations (which assess the extent to which a program achieves its objectives), and
- Cost-benefit and cost-effectiveness analyses (which assess a program's outputs or outcomes with the costs to produce them).[1]

Deciding which evaluation type to use involves a variety of different considerations, as no one evaluation is suitable for all programs. For instance, as we have previously reported, an impact evaluation is more likely to provide useful information about what works when the intervention consists of clearly defined activities and goals and has been well implemented.[2] One type of impact evaluation—the quasi-experimental comparison group design—which compares outcomes for program participants with those of a similar group not in the program, is used in instances when random assignment to the participant and nonparticipant groups is not possible, ethical, or practical. It is most successful in providing credible estimates of program effectiveness when the groups are formed in parallel ways and are not based on self-selection. On the other hand, case studies are recommended for assessing the effectiveness of complex interventions in limited circumstances when assessing comprehensive reforms that are so deeply integrated with the context (for example, the community) that no truly adequate comparison case can be found. Furthermore, every research method has inherent limitations; therefore, it is often advantageous to combine multiple measures or two or more designs in a study or group of studies to obtain a more comprehensive picture of the program's effect.

As we have also previously reported, the evaluation methods literature describes a variety of issues to consider in planning which methods to use in carrying out an evaluation, including the expected use of the evaluation, the nature and implementation of program activities, and the resources available for the evaluation. We identified the following methods and designs of evaluation in our review, which may be used to carry out one or more of the main types of evaluation listed above:

- committee of visitors, and other report types, which are generally external peer reviews that examine programs' managerial stewardship, compare plans with progress made, and evaluate outcomes to determine whether the research contributes to the agency's mission and goals;
- experimental methods, which involve randomly assigning one group to a program and another to not participate in the program in order to compare outcomes of both groups;
- mixed methods, which combine qualitative and quantitative designs;
- qualitative, such as interviews or focus groups;
- surveys, which involve the systematic collection of data from a respondent using a structured instrument (i.e., a questionnaire) to ensure that the collected data are as accurate as possible; and
- quasi-experimental comparison groups.

In addition, there were two evaluations based solely on a compilation of grantee reports. As stated previously, other evaluations also used grantee evaluations, but these used other data sources to inform their results, and so were classified as using either mixed or qualitative methods. The most common evaluation designs that we classified programs as using were the committee of visitors and mixed methods.

We reviewed 35 evaluations from the following agencies and programs, and determined their primary method for assessing effectiveness:

Table 4. Program Evaluations and Evaluation Methods

Agency	Program	Evaluation	Methods
Department of Commerce, National Institute of Standards and Technology	Summer Institute for Middle School Science Teachers	Evaluation of the National Institute of Standards and Technology's (NIST) Summer Institute Year 3 Report	Mixed
Department of Commerce, National Institute of Standards and Technology	Summer Undergraduate Research Fellowship Program	NIST SURF Program Assessment	Survey
Agency	Program	Evaluation	Methods
Department of Commerce, National Oceanic and Atmospheric Administration	Bay Watershed Education and Training Program (B-WET)	An Evaluation of National Oceanic and Atmospheric Administration Chesapeake Bay Watershed Education and Training Program Meaningful Watershed Educational Experiences	Mixed
Department of Commerce, National Oceanic and	Global Learning and Observations to Benefit the Environment	GLOBE 10 Year Evaluation: Into the Next Generation	Quasi-experimental
Atmospheric Administration	(GLOBE) Program		
Department of Commerce, National Oceanic and Atmospheric Administration	Teacher at Sea Program	Evaluation Report of the NOAA Teacher at Sea Program: 2005-2009	Mixed

Agency	Program	Evaluation	Methods
Department of Defense	National Defense Education Program (NDEP) K-12	Recommended Resources for the National Defense Education Program Pre-Engineering Partnerships	Qualitative
Department of Defense	DoD STARBASE Program	DoD STARBASE Program: 2010 Annual Report	Mixed
Department of Defense	Army Educational Outreach Program (AEOP)	The Talent Imperative in Science and Technology: An Evaluation of Army Educational Outreach Programs	Mixed
Department of Education	Graduate Assistance in Areas of National Need	A Study of Four Federal Graduate Fellowship Programs: Education and Employment Outcomes	Survey
Department of Education	Mathematics and Science Partnerships	Mathematics and Science Partnerships: Summary of Performance Period 2008 Annual Reports–Analytic and Technical Support for Mathematics and Science Partnerships	Analysis of grantee evaluations
Department of Education	National Science and Mathematics Access to Retain Talent Program	Academic Competitiveness and National SMART Grant Programs: 2006-07 and 2007-08	Mixed
Department of Education	Upward Bound Math-Science	The Impacts of Upward Bound Math-Science on Postsecondary Outcomes 7–9 Years After Scheduled High School Graduation: Final Report	Quasi-experimental
Department of Energya	Science Undergraduate Laboratory Internships	Making Comparisons and Examining Experiences: A Program Evaluation of the Department of Energy's Student Undergraduate Laboratory Internship (SULI) Program	Mixed
Department of Health and Human Services, National Institutes of Health	Curriculum Supplement Series	The Relative Effects and Equity of Inquiry-Based and Commonplace Science Teaching on Students' Knowledge, Reasoning, and Argumentation	Experimental
National Institutes of Health, Department of Health and Human Services	NIH Summer Research Experience Programs	Stimulating Science Education: NIH Summer Research Program Engages Students and Teachers in Science	Survey
National Institutes of Health, Department of Health and Human Services	NIH Undergraduate Scholarship Program for Individuals from Disadvantaged Backgrounds	The NIH Undergraduate Scholarship Program: Career Outcomes of Scholars and Non-Awarded Finalists	Quasi-experimental
Environmental Protection Agency	National Environmental Education and Training Partnership	The Third Environmental Education and Training Partnership: Summary of Year 5 Achievements	Analysis of grantee evaluations

Table 4. (Continued)

Agency	Program	Evaluation	Methods
Environmental Protection Agency	EPA Greater Research Opportunities (GRO) Fellowships for Undergraduate Environmental Study	Review of the Office of Research and Development's Science to Achieve Results (STAR) and Greater Research Opportunities (GRO) Fellowship Programs at the U.S. Environmental Protection Agency	Mixed
Environmental Protection Agency	Science to Achieve Results Graduate Fellowship Program	Review of the Office of Research and Development's Science to Achieve Results (STAR) and Greater Research Opportunities (GRO) Fellowship Programs at the U.S. Environmental Protection Agency	Mixed
NASA	NASA Informal Education Opportunities (NIEO)	NASA Informal Education: Final Report—A Descriptive Analysis of NASA's Informal Education Portfolio: Preliminary Case Studies	Mixed
NASA	Space Grant/EPSCoR Program	20th Year Program Evaluation Executive Summary: National Space Grant College and Fellowship Program	Mixed
National Science Foundation	Alliances for Graduate Education and the Professoriate (AGEP)	National Evaluation of the Alliances for Graduate Education and the Professoriate	Mixed
National Science Foundation	CISE Pathways to Revitalized Undergraduate Computing Education (CPATH)	Evaluation of CISE Pathways to Revitalized Undergraduate Computer Education (CPATH)	Mixed
National Science Foundation	Engineering Education (EE)	Early Outcomes of the National Science Foundation's Grants Program on "How People Learn Engineering" (HPLE)	Qualitative
National Science Foundation	Federal Cyber Service: Scholarship for Service	Federal Cyber Service: Scholarship for Service Program Summative Evaluation Report	Mixed
National Science Foundation	Integrative Graduate Education and Research Traineeship (IGERT) Program	Evaluation of the National Science Foundation's Integrative Graduate Education and Research Traineeship Program (IGERT): Follow-up Study of IGERT Graduates: Final Report	Quasi-experimental
National Science Foundation	Louis Stokes Alliances for Minority Participation (LSAMP)	Final Report on the Evaluation of the National Science Foundation Louis Stokes Alliances for Minority Participation Program	Quasi-experimental

Agency	Program	Evaluation	Methods
National Science Foundation	Research Experiences for Teachers (RET) in Engineering and Computer Science	Evaluation of the Research Experiences for Teachers (RET) Program: 2001-2006 Final Report	Survey
National Science Foundation	Research Experiences for Undergraduates (REU)	A Draft Report to the National Science Foundation: Research Experiences for Undergraduates (REU) in the Directorate for Engineering (ENG): 2003-2006 Participant Survey	Survey
National Science Foundation	Research in Disabilities Education (RDE)	Research in Disabilities Education Program Evaluation: Study 1 Methods and Results	Mixed
National Science Foundation	Robert Noyce Teacher Scholarship Program	Robert Noyce Teacher Scholarship Program: Synopsis of 5 Years of Evaluation	Mixed
National Science Foundation	Graduate STEM Fellows in K-12 Education Program (GK-12)	Evaluation of the National Science Foundation's GK-12 Program: Final Report, Volumes I and II: Technical Report and Appendices	Quasi-experimental
National Science Foundation	The Historically Black Colleges and Universities Undergraduate Program (HBCU-UP)	Capacity Building to Diversify STEM Realizing Potential Among HBCUs: Findings from the National Evaluation of the Historically Black Colleges and Universities Undergraduate Program	Quasi-experimental
National Science Foundation	Enhancing the Mathematical Sciences Workforce in the 21st Century (EMSW21)	Evaluation of NSF's Program of Grants and Vertical Integration of Research and Education in the Mathematical Sciences (VIGRE)	Mixed
Department of Agriculture	1890 Institution Teaching and Research Capacity Building Grants Program	Portfolio Annual Report 2008: Education	Other

Source: GAO analysis of survey results.

[a]Department of Energy officials also submitted a committee of visitors report, which we list below in table 5.

The following are different types of reports, including the committee of visitors, that programs used to assess effectiveness of their STEM education programs. As stated in appendix I, we did consider these to be evaluations but did not review them because they did not align with the criteria we used to assess the evaluations.

Table 5. Committee of Visitors and Other Types of Reports Used to Assess Program Effectiveness

Agency	Program	Evaluation
Department of Defense	Uniformed Services University of the Health Sciences (USUHS)	Review of the School of Medicine
Department of Energy	Academies Creating Teacher Scientists (DOE Acts)	Committee of visitors

Table 5. (Continued)

Agency	Program	Evaluation
Department of Energy	American Chemical Society Summer School in Nuclear and Radiochemistry	Peer-reviewed report
Department of Energy	ASCR-ORNL Research Alliance in Math and Science	Internal report
Department of Energy	Community College Institute of Science and Technology	Committee of visitors
Department of Energy	Faculty and Student Teams	Committee of visitors
Department of Energy	Hampton University Graduate Studies	Peer-reviewed report
Department of Energy	Laboratory Equipment Donation Program	Committee of visitors
Department of Energy	National Science Bowl	Committee of visitors
Department of Energy	National Undergraduate Fellowship Program in Plasma Physics and Fusion Energy Sciences	Internal report
Department of Energy	Office of Science Graduate Fellowship (SCGF) Program	Committee of visitors
Department of Energy	Pan American Advanced Studies Institute	(Program officials did not provide us the name of this evaluation, but did note that it is jointly supported with NSF and NSF led the program's peer review.)
Department of Energy	Plasma/Fusion Science Educator Programs	Committee of visitors
Department of Energy	Pre-Service Teacher Program	Committee of visitors
Department of Energy	Science Undergraduate Laboratory Internships	Committee of visitors
Department of Energy	Summer Applied Geophysical Experience (SAGE)	Peer-reviewed report
National Institutes of Health, Department of Health and Human Services	Ruth L. Kirschstein National Research Service Award Institutional Research Training Grants (T32, T35)	Online performance index
National Science Foundation	Advanced Technological Education (ATE)	Committee of visitors
National Science Foundation	Discovery Research K-12 (DR-K12)	Committee of visitors
National Science Foundation	Ethics Education in Science and Engineering (EESE)	Committee of visitors
National Science Foundation	Geoscience Education	Committee of visitors
National Science Foundation	Geoscience Teacher Training (GEO-Teach)	Committee of visitors
National Science Foundation	Global Learning and Observations to Benefit the Environment (GLOBE)	Committee of visitors
National Science Foundation	Graduate Research Fellowship Program (GRFP)	Committee of visitors
National Science Foundation	Opportunities for Enhancing Diversity in the Geosciences	Committee of visitors
National Science Foundation	Transforming Undergraduate Education in STEM (TUES)	Committee of visitors
National Science Foundation	Undergraduate Research and Mentoring in the Biological Sciences (URM)	Committee of visitors

Source: GAO analysis of survey results.

End Notes

[1] GAO, Higher Education: Federal Science, Technology, Engineering, and Mathematics Programs and Related Trends, GAO-06-114 (Washington, D.C.: Oct. 12, 2005).

[2] U.S. Department of Education, Report of the Academic Competitiveness Council, Washington, D.C., 2007.

[3] See appendix I for our definition of a STEM education program.

[4] In GAO-06-114, we counted three organizations—National Institutes of Health (NIH), Indian Health Service (IHS), and Health Resources and Services Administration—within the Department of Health and Human Services (HHS) as three separate agencies. In this report, we count all subagencies, agencies, and organizations of cabinet-level departments as one agency. Therefore, NIH and IHS are counted as one agency—HHS— in this report.

[5] President's Council of Advisors on Science and Technology; Report to the President: Prepare and Inspire: K-12 Education in Science, Technology, Engineering, and Math (STEM) for America's Future. Washington, D.C., September 2010.

[6] Exec. Order No. 12881 (1993).

[7] STEM programs may fall under the jurisdiction of several congressional committees including those that oversee science, space, and technology programs; defense and homeland security programs; and education programs.

[8] Pub. L. No. 110-315, 122 Stat. 3078 (2008).

[9] Pub. L. No. 107-110, 115 Stat. 1425 (2002), reauthorizing and amending the Elementary and Secondary Education Act of 1965.

[10] Pub. L. No. 81-507, 64 Stat. 149.

[11] Pub. L. No. 109-171, tit. VIII, § 8003, 120 Stat. 4, 155 (2006).

[12] Pub. L. No. 110-69, 121 Stat. 572 (2007). COMPETES also focused on STEM research programs.

[13] America COMPETES Reauthorization Act of 2010, Pub. L. No. 111-358, 124 Stat. 3982.

[14] Pub. L. No. 111-358, § 101, 124 Stat. 3982, 3984.

[15] Pub. L. No. 111-352, 124 Stat. 3866.

[16] The GPRA Modernization Act of 2010 uses the term "agency," which is defined as an executive department, a government corporation, or an independent establishment, but does not include the Central Intelligence Agency, the Government Accountability Office, the U.S. Postal Service, and the Postal Regulatory Commission. 5 U.S.C. § 306(f).

[17] Pub. L. No. 111-139, § 21, 124 Stat. 8, 29 (2010), codified at 31 U.S.C. § 712 note.

[18] GAO, Opportunities to Reduce Potential Duplication in Government Programs, Save Tax Dollars, and Enhance Revenue, GAO-11-318SP (Washington, D.C.: Mar. 1, 2011). An interactive, web-based version of the report is available at: http://www.gao.gov/ereport/gao-11-318SP.

[19] Our analysis of programs does not include the 29 earmarks that were funded in 2010 because, according to our survey, 25 of these were not funded in 2011.

[20] Nine program officials indicated that they did not know whether the program was created under their agencies' general statutory authority or through congressional direction.

[21] GAO asked survey respondents to report on obligations—defined as definite commitments that create a legal liability of the government for the payment of goods and services ordered or received, or a legal duty on the part of the United States that could mature into a legal liability.

[22] These other federal efforts are not included in our analysis of federal STEM education programs because the primary objective is not STEM education, but a secondary or tertiary benefit would be to enhance STEM education. Examples in this section are not intended to be exhaustive, but rather to provide examples of the different types of activities that can contribute to furthering STEM education and competitiveness. Further, many agencies administer efforts to promote STEM employment, such as funding postdoctoral students to perform research, which are not included in our review.

[23] T hrough the GeoFORCE program, Interior has provided non-financial resources such as speakers, experts, science information materials, and mentoring. The agency has also provided the program a small amount of financial assistance intermittently; however, the main contribution has been non-financial.

[24] African-Americans and Hispanics or Latinos were the most frequently identified minority, disadvantaged, or underrepresented groups.

[25] Committee on STEM Education. National Science and Technology Council. Executive Office of the President of the United States. The Federal Science, Technology, Engineering, and Mathematics (STEM) Education Portfolio. A Report from the Federal Inventory of STEM Education Fast-Track Action Committee. Washington, D.C.: December 2011.

[26] GAO-11-318P.

[27] Not all programs use the same definition of administrative costs, and programs reported various methods to develop estimated costs.

[28] For more details on our review of agencies' annual performance plans and reports, see appendix I.

[29] In addition to reporting on STEM education programs through their performance plans and performance reports, there may be other ways to report on these efforts. However, our analysis was limited to these two documents. See appendix I for more information on our scope and methodology.

[30] For more details on our evaluation review, see appendix I and appendix III.

[31] Four of the evaluations did not have enough information for us to make a determination of the extent to which the methods, questions, objectives, and program context were aligned.

[32] The National Science and Technology Council Committee on Science Subcommittee on Education and Workforce. Review and Appraisal of the Federal Investment in STEM Education Research. October 2006.

End Notes for Appendix I

[1] GAO-06-114.

[2] U.S. Department of Education, Report of the Academic Competitiveness Council, Washington, D.C., 2007.

[3] President's Council of Advisors on Science and Technology. Report to the President: Prepare and Inspire: K-12 Education in Science, Technology, Engineering, and Math (STEM) for America's Future. Washington, D.C., September 2010.

[4] National Academies of Science, National Academy of Engineering, and Institute of Medicine of the National Academies, Rising above the Gathering Storm: Energizing and Employing America for a Brighter Economic Future: Committee on Prospering in the Global Economy of the 21stCentury: An Agenda for American Science and Technology. National Academies Press, 2007.

[5] Committee on STEM Education. National Science and Technology Council. Executive Office of the President of the United States. The Federal Science, Technology, Engineering, and Mathematics (STEM) Education Portfolio. A Report from the Federal Inventory of STEM Education Fast-Track Action Committee. Washington, D.C., December 2011.

[6] To develop these objectives, we first examined the objectives used by previous GAO work to inventory federal STEM education programs. Second, we discussed these objectives with agency officials; the discussion resulted in providing clarifying language for some objectives and adding a new objective on conducting research to enhance the quality of STEM education. Third, after review and analysis of survey responses, we determined that for purposes of reporting out on survey responses, we would combine the first four objectives into one broader objective category—attracting and preparing students throughout their academic careers to enter STEM fields.

[7] Although we surveyed 29 earmarks that were funded in 2010, we did not include earmark data in our analysis because, according to our survey, 25 of these were not funded in 2011.

[8] After initial deployment of our survey, we became aware of four new programs not previously on our list—the National Security Agency's Cryptanalysis and Exploitation Services Summer Program, the National Institutes of Health's Material Development for Environmental Health Curriculum, the Animal and Plant Health Inspection Service's Daniel E. Salmon Scholarship, and the AgDiscovery program. After speaking with program officials and reviewing program information, we determined that these programs met our definition of a STEM education program, so we obtained program contact information, had each program fill out a survey, and added them to our review. In addition, there were five programs that program officials said should be excluded from our review after receiving the survey, even though agency officials had confirmed the list at the outset. After speaking with officials and reviewing program information, we determined that all five programs should be excluded from our list and should not fill out the survey. It was determined that two programs were part of another program, one was a duplicate entry, one was a nonprogrammatic STEM activity, and one was a research-related program not focused on STEM education.

[9] GAO-11-646SP.

[10] We did not assess agencies' plans and reports for compliance with GPRA and the Government Performance and Results Modernization Act of 2010 requirements, and our findings that some agencies did not include STEM education programs in their plans and reports should not be read to suggest that we identified instances of noncompliance. For example, we did not assess whether a particular STEM education program is a "program

activity" as that term is defined by GPRA for purposes of determining what STEM education programs are required to be covered in agency performance plans and reports. 31 U.S.C. § 1115(h)(11).

End Notes for Appendix III

[1] GAO-11-646SP.
[2] GAO-10-30.

INDEX

D

E

F

G

H

I

J

K

L

S

T

U

V

W

Y